THE CHARACTER OF CONSENT

Information Policy Series

Edited by Sandra Braman

The Information Policy Series publishes research on and analysis of significant problems in the field of information policy, including decisions and practices that enable or constrain information, communication, and culture irrespective of the legal siloes in which they have traditionally been located as well as state-law-society interactions. Defining information policy as all laws, regulations, and decision-making principles that affect any form of information creation, processing, flows, and use, the series includes attention to the formal decisions, decision-making processes, and entities of government; the formal and informal decisions, decision-making processes, and entities of private and public sector agents capable of constitutive effects on the nature of society; and the cultural habits and predispositions of governmentality that support and sustain government and governance. The parametric functions of information policy at the boundaries of social, informational, and technological systems are of global importance because they provide the context for all communications, interactions, and social processes.

A complete list of the books in the Information Policy series appears at the back of this book.

THE CHARACTER OF CONSENT

THE HISTORY OF COOKIES AND THE FUTURE OF TECHNOLOGY POLICY

MEG LETA JONES

The MIT Press
Cambridge, Massachusetts
London, England

© 2024 Massachusetts Institute of Technology

All rights reserved. No part of this book may be used to train artificial intelligence systems or reproduced in any form by any electronic or mechanical means (including photocopying, recording, or information storage and retrieval) without permission in writing from the publisher.

The MIT Press would like to thank the anonymous peer reviewers who provided comments on drafts of this book. The generous work of academic experts is essential for establishing the authority and quality of our publications. We acknowledge with gratitude the contributions of these otherwise uncredited readers.

This book was set in ITC Stone Serif Std and ITC Stone Sans Std by New Best-set Typesetters Ltd. Printed and bound in the United States of America.

Library of Congress Cataloging-in-Publication Data

Names: Jones, Meg Leta, author.
Title: The character of consent : the history of cookies and the future of technology policy / Meg Leta Jones.
Description: Cambridge, Massachusetts : The MIT Press, 2024. | Series: Information policy series | Includes bibliographical references and index.
Identifiers: LCCN 2023028059 (print) | LCCN 2023028060 (ebook) |
 ISBN 9780262547949 (paperback) | ISBN 9780262378451 (epub) |
 ISBN 9780262378444 (pdf)
Subjects: LCSH: Cookies (Computer science)—Social aspects. | Cookies (Computer science)—History. | Data privacy. | Consent (Law) | Technology and state.
Classification: LCC QA76.9.C72 J66 2024 (print) | LCC QA76.9.C72 (ebook) |
 DDC 006.7—dc23/eng/20240125
LC record available at https://lccn.loc.gov/2023028059
LC ebook record available at https://lccn.loc.gov/2023028060

10 9 8 7 6 5 4 3 2 1

CONTENTS

SERIES EDITOR'S INTRODUCTION   vii

1   INTRODUCTION: COOKIES AND CONSENTING CHARACTERS   1

2   COMPUTING THE DATA SUBJECT   25

3   NETWORKING THE USER   73

4   MAINTAINING STATE FOR THE PRIVACY CONSUMER   105

5   CONTESTING COOKIES FOR DATA PRIVACY   141

6   CONCLUSION: THE (FORCED) RETIREMENT OF COOKIES   169

NOTES   191
INDEX   261

SERIES EDITOR'S INTRODUCTION

SANDRA BRAMAN

It is not that we go looking for information—when we are online, information comes looking for us. Meg Leta Jones uses the transactional point at which that happens, the cookie, as her lens on the multiple histories of privacy law with an eye to what they mean for social relations and for governance. The concept of the "sociotechnical" has come into heavy use but as an intellectual practice it is so rarely genuinely achieved; in *The Character of Consent*, Jones does so in spades. She provides insights into the different kinds of human beings, or characters, who are created, revealed, or enabled by different approaches to privacy in the digital environment. She does so through an analysis that contrasts American and European privacy law and looks across time, beginning with the origins of the commercial computer industry after World War II. The combination of two kinds of intellectual contributions—clear stories and impressively detailed historical information from an unusually broad range of sources—makes this the rare book that lets us see both forest and trees. *The Character of Consent* is a book for the short shelf on privacy you keep nearby.

Jones makes good use of a range of resources that include archives, legal documents, public discourse, interviews with key actors, and more, to identify idiosyncratic, culturally based, economically driven, political, and structural factors shaping particular privacy protection efforts. Key but often forgotten developments are recuperated, as in the stories she tells about what many appreciate as the original "Silicon Valley"—the Twin Cities of Minneapolis and St. Paul, Minnesota, home to the international headquarters of IBM and other industry-shaping mainframe computer companies. In other areas, the information was always obscure, but life is better knowing that Boy Scouts who hacked Los Alamos spurred passage of the Computer Fraud and Abuse Act (CFAA), a key element of US cybersecurity law. Jones has a taste for the delightful detail, reporting that it wasn't until 1971 that the British Computer Society approved its code of conduct . . . a year after it put in place its official coat of arms.

Although this is not a book about gender and technology, it is a particularly powerful book about women and technology. Jones achieves this by simply making sure the details are there. Thus, we learn that it wasn't Tim Berners-Lee but Nicola Pellow, an undergraduate math intern working at CERN, who first wrote code for the web that was not system dependent. It was not Marc Andreessen but Colleen Bushell, trained in visualization by Donna Cox, who created the interface design for the first version of Mosaic. Once Mosaic turned into Netscape, it was again not Andreessen but Rosanne Siino, the company's vice president of communications, who envisioned using the service to lure everyone onto the web. Wenda Harris Millard gets the credit for developing online advertising. When, after World War II, the British government tried to replace the women who had run the computer industry during the war with men, it damaged the industry itself in ways that had long-term competitive effects. And it wasn't Mark Amerika who wrote the first set of cookies that made it possible to alter readers' paths through his digital artwork, it was an undergraduate at Brown—danah boyd.

Over and over, Jones reports on the irrepressible introduction of new privacy problems with each technological innovation, and on the experimentation with new ways of attempting to protect privacy that inevitably follows. The cookie as an internet concept first appeared in April 1987 in a technical internet design document, RFC 1004, "Distributed-Protocol Authentication Scheme." What that academic author, D. L. Mills, proposed was somewhat different from the kind of cookies Jones discusses, starting with the Netscape version introduced a decade later, but the family relation is clear. The problem addressed in RFC 1004 was the need for trust and threat protection. The solution was a "cookie jar," to be provided by an authentication server that would "permit only those associations sanctioned by the Cookie Jar while operating over arbitrary network topologies, including non-secured networks and broadcast-media networks, and in the presence of hostile attackers."[1] The document emphasized that the purpose was *not* to protect the privacy of what was being transmitted other than private keys.

From Jones we learn that when the French government first confronted digital privacy issues in the 1970s, it prohibited decision-making on the basis of computerized processing of personal information and there was a right to contest not only the data, but the logic(s) used to process that data. There was a British suggestion that people have a "right to a print-out." This book's history of the many variants of legal digital privacy protections in the US and the EU, whether or not ever tried, and whether or not ever successful if they were, can be mined as a rich source of ideas for governance strategies for today's environment. The timing is good, as everything is under reconsideration, beginning with basic information policy principles. The EU's "right to be forgotten" is now familiar but, in terms of human history, is quite new. Other new principles, such as a state's "right *not* to know," are less discussed and not necessarily acknowledged even though they are de facto in implementation.

This book on cookies is fecund regarding future research trajectories. Analysis of privacy in cultures other than those discussed here must inevitably reveal additional characters. The Warlpiri with whom ethnographer Eric Michaels lived, for example, as reported on in his extraordinary *Bad Aboriginal Art*, define privacy boundaries relationally. Neither space, nor audibility, nor visibility are relevant. It is interpersonal relationships that create informational structures, with death as a significant barrier—it is forbidden to view the photograph of anyone who has died. Who are the privacy characters here?

In another example of research questions this book opens up, Jones reports that punch cards remained in use into the 1980s, long after electronic storage media were also widely in use, because they could be changed without a computer. The American insurance industry also stayed with ammonia replicators, despite their health cost, long after other industries went to mimeograph and ditto. Letterpress printing didn't die when offset came along; it turned to the service of highly prized book art and craft. Research into why purportedly obsolete or legacy technologies continue to be used, and the consequences of such practices, would be worthy of a book of its own.

The Character of Consent stands out among books on privacy in that it is not just about the law and its social effects but about the law's intertwining with the development of the computer industry and with changes in the nature of the state. The relationship between consent mechanisms and larger political endeavors is non-trivial, as Jones's title emphasizes. Meanwhile, those involved with the ongoing technical design of the internet are still thinking about cookies. There are eleven RFCs so far dealing with cookies. The most recent was RFC 9018, "Interoperable Domain Name System (DNS) Server Cookies," with four authors, two from corporate technology producers and two from the Internet Systems Consortium.[2] It was published in April 2021—after Google made its announcement that it was turning away from the use of cookies and other browsers had already taken that path. Tensions between geopolitical and network political governance remain.

1 INTRODUCTION: COOKIES AND CONSENTING CHARACTERS

If you grabbed lunch on the run in downtown DC circa 2001, you might have discovered a free bag of desserts asking, "Do you know where your cookies come from?"[1] The cookies were a promotion for the Atlanta-based internet service provider EarthLink, which had been trying to sell itself as a privacy-friendly alternative to America Online using everything from television ads to toilet seat covers.[2] As its vice president of brand marketing put it, "Our position is you should be able to understand what's being revealed about you, and you should be able to control it."[3] Fast-forward twenty years and a *Wired* magazine headline reads, "How to Avoid Those Infuriating Cookie Pop-Ups."[4] The article complains that the most recent European law "was meant to make it easier for people to understand and control how they are tracked online. In reality, it has made the internet even more unusable." Confronting cookies remains one of the most quintessential technological experiences of daily life, decades after their introduction to the first mainstream web browser.

A cookie is a string of data (e.g., GA1.2.998356558.GHIKWWPO) generated by the site being visited and sent to the individual's browser, where it is then stored on the individual's hard drive in the web browser folder. When an individual visits a site for the first time, the web's hypertext transfer protocol (HTTP) can include a Set-Cookie response header, which uses "name-value" pairs. So

```
Set-Cookie: <cookie-name>=<cookie-value>;
Expires=<date> may generate Set-Cookie:
id=sf2kjasfe3gp0ig; Expires=Sun, 14 Feb 2021 12:00:00
EST.
```

When an individual revisits or further explores the site, the browser automatically sends the cookie back to the server and tells the server what data to recall. Indiscernible to the individual requester, the cookie is an "opaque" memory system.[5] This is how the web has "remembered" since the mid-1990s, when Netscape Navigator added the feature to its immensely popular browser. Histories of the term "cookie," the design of the memory system, and the browser will be explored in detail in the following pages.

Cookies have since had quite a run, not least because they are so flexible. Entering a URL into the browser address bar is a request for content from www.website.com's server, but that content may be assembled with material, like a picture or recent social media posts, pulled from another server and inserted into a designated area of the web page. This is an opportunity for another server to set a cookie—a third-party cookie. The number of cookies set by the first party (the domain in the request) and third parties varies dramatically and has changed over time as advertising players have changed. In 2010, the *Wall Street Journal*'s groundbreaking "What They Know" series reported that over half of the top fifty websites in the US installed at least twenty-three third-party cookies, and that Dictionary.com installed 159.[6] A 2016 study crawled the top 100,000 most popular websites and found at least one cookie on 94 percent of sites, up from 67 percent reported in a 2009 study conducted using a similar crawling method.[7] Of the 100,000 websites, an average of nineteen cookies were set per site, and over twenty sites set over 200 cookies each.[8] In 2020, the HTTP Archive tracked around 5.8 million home pages: 84 percent used cookies and 79 percent of those were third-party cookies.[9]

All these cookies don't actually infuriate people—their accompanying cookie notifications, click-throughs, and pop-ups infuriate people. At the beginning of the 2020s, people everywhere use personal computer screens large and small to interact with websites from around the world, but this often requires a few frustrating maneuvers to select or cancel a box that pops up describing the way sites and services collect and use cookies. Weary internet travelers click through and dodge endless cookie notifications each day (see figure 1.1). Cookies are a nuisance, because they interfere with the user experience on sites and platforms and represent a political failure to provide meaningful privacy protection. Notification practices like privacy policies and terms of service have justified the extensive use of cookies as part of an "arrangement" wherein services are exchanged for

INTRODUCTION 3

FIGURE 1.1
Common cookie consent interface.

data. Individuals "agree" to the arrangement by continuing to the site or clicking a box, thereby "consenting" to invasive data collection, analysis, and sharing. This phony performance of privacy based on consent—what science fiction novelist and tech journalist Cory Doctorow calls "consent theater"—has plagued networked computer communications for decades.[10]

How is the phenomenon of consent theater explained? How did a technology that was originally designed to promote additional functionality, privacy, and decentralized websites become a tool that is essential to an invasive, choiceless, and clunky internet? When Congress asked Mark Zuckerberg these questions in 2018, he answered:

> I think we and others in the tech industry have found [it] challenging, which is that long privacy policies are very confusing. And if you make it long and spell out all the detail, then you're probably going to reduce the percent of people who read it and make it accessible to them. So, one of the things that—that we've struggled with over time is to make something that is as simple as possible so people can understand it, as well as giving them controls in line in the product in the context of when they're trying to actually use them, taking into account that we don't expect that most people will want to go through and read a full legal document.[11]

His response is not far off from scholarly accounts of consent. Privacy researchers and thinkers describe the contemporary consent arrangement as the product of a system broken by the flood of digital information and assert that the existing consent model cannot protect people in the face of the sheer breadth of the internet. These accounts focus on the challenges of self-managing our own privacy: namely, that our ability to provide meaningful consent to tracking is limited by the quantity and quality of notifications, as well as the lack of alternatives.[12] Privacy researchers have shown that we'd all have to take a month off work to read all the privacy policies we encounter each year.[13] Privacy policies are written in complex legalese in order to convey complicated information processing and practices (and even then, rarely provide specific details about whose information is shared with whom and how it might be used or further shared).[14] Even if it were reasonable to read and understand all these policies, there isn't usually much point. Privacy policies are not negotiable; they are presented in take-it-or-leave-it terms.[15] Choosing an alternative service is often unrealistic, if not impossible.[16]

These privacy self-management woes have been referred to as the "consent dilemma," which is described as such because consent has been a central moral and legal tenet of contemporary digital governance.[17] Its centrality is unsurprising in our long era of lionized consent.[18] Given the cultural and contextual richness required to grapple with consent, it is also unsurprising that its extensive use as a governance tool has led to confusion and inconsistencies. Stated broadly, an individual's consent involves an effective communication of an intentional transfer of rights and obligations between parties.[19] As a moral concept, consent transforms the relationship between two parties—it is "moral magic."[20] Consent is also legal magic. Valid consent can transform sexual assault, theft, battery, and an invasion of privacy into sexual relations, borrowing a car, surgery, and the legitimate use of personal information. The magic of consent lies in its ability to render permissible otherwise impermissible actions.

Framing privacy as a challenge of self-management poses the problem as one of magnitude that can be solved by streamlining. There's something unsatisfying about this solution, because privacy scholars and engineers have been so successful at meeting the demands of the consent dilemma. Technical tools to manage cookies and ideas for simplifying consent were produced almost immediately. Users have been blocking cookies for as long as they have been aware of them. In 1996, the first *Wired* magazine article about cookies covered a program called PGPcookie.cutter that blocked or managed cookies and could be downloaded for $29.95.[21] A robust body of research has developed to create systems and standards to support informed consent and choice in digital environments. And yet, remarkably, cookies, cookie banners, and click-throughs *still* plague user experiences and web functionality at the time of writing. I argue, however, that the consent dilemma is not an information overload crisis, but an identity crisis.

Instead of asking how people consent (or not), I ask *who* exactly is supposed to be consenting to *what* and *why*. Posing such a simple question about cookies has a surprisingly complicated answer, which makes cookies an ideal issue for political mischief. Cookies and their attendant consent mechanisms are actually regulated by three areas of law: data protection, communication privacy, and consumer protection. Tracing them separately provides an alternative avenue for understanding the consent dilemma and possible ways to frame future tech policy issues. To do so, I follow the *who* at the heart of each.

MAKING UP COMPUTER CHARACTERS

In the same 2018 congressional hearing, Mark Zuckerberg pulled off a remarkable feat. He received question after question from Congress about how Facebook could have exposed the data of approximately 87 million users without their consent during the Cambridge Analytica data scandal.[22] Somehow, over the course of two days, Zuckerberg never once uttered the word *user*, despite congresspeople using the term hundreds of times. In each instance, he carefully adjusted, as you can see in the emphasis added in the following exchanges.

When asked, "And the privacy settings, it's my understanding that they limit the sharing of that data with other Facebook *users*, is that correct?"

Zuckerberg responded, "Senator, yes. Every *person* gets to control who gets to see their content."

"Why doesn't Facebook disclose to its *users* all the ways that data might be used by Facebook and other third parties?"

"Mr. Chairman, I believe it's important to tell *people* exactly how the information that they share on Facebook is going to be used."

"But do you collect *user* data through cross-device tracking?"

"Senator, I believe we do link *people's* accounts between devices in order to make sure that their Facebook and Instagram and their other experiences can be synced between their devices."

At various points, congresspeople referred to *consumers* instead. One senator stated, "The status quo no longer works. Moreover, Congress must determine if and how we need to strengthen privacy standards to ensure transparency and understanding for the billions of *consumers* who utilize these products." Another argued, "To my mind, the issue here is transparency. It's *consumer* choice. Do *users* understand what they're agreeing to—when they access a website or agree to terms of service? Are websites up-front about how they extract value from *users*, or do they hide the ball? Do *consumers* have the information they need to make an informed choice regarding whether or not to visit a particular website?" But Zuckerberg carefully avoided calling his Facebook friends *consumers*. Slipping only once when asked what kind of legislation he thought would be helpful, Zuckerberg explained that "[Privacy policies] need to be implemented in a way where people can actually understand it, where *consumers* can—can understand it, but that can also capture all the nuances of how these services work in a way that doesn't—that's not overly restrictive on—on providing the services."

Why would Zuckerberg go to so much trouble?

User and *consumer* are loaded terms. Zuckerberg himself is a user and a consumer. When Silicon Valley leaders are criticized for being unfairly unregulated, they point out that they adhere to tax codes, employment laws, municipal ordinances, and so on around the world. In those legal contexts, Zuckerberg is a taxpayer, employer, and property owner. Each conjures a unique relationship to a particular area of law, rights, and expectations with different societal goals and different cultural associations. So too do users and consumers.

This book tells the history of digital consent through the familiar cookie by following the people who are supposed to be consenting. It presents the history as a story of three characters: the *data subject*, the *user*, and the *privacy consumer*. All three are made-up people. In 1986, Ian Hacking published "Making Up People" and later argued that "new scientific classification may bring into being a new kind of person, conceived of and experienced as a way to be a person."[23] For decades, Hacking has analyzed different made-up people. For instance, he has analyzed the evolution of the "child abuser" since the 1960s, when battered child syndrome was first discussed, and the later developments of "pedophile," "stranger danger," and "sex offender."[24] Hacking's explorations of classifications like "poverty line" and "body mass index" and classes like "poor" and "obese" highlight how they "conspire to emerge hand in hand, each egging the other on."[25] Using this theoretical basis, I explore the data subject, the user, and the privacy consumer as made-up *computer characters*—all studiously avoided by Zuckerberg.[26]

These characters come from various technology governing processes that, as with Hacking's cases, make people up. The computer characters are classifications of people in relation to computers produced in different types of legal documents like legislation, regulation, agency reports, court cases, and international treaties and directives, all with particular forms of influence from technical communities. Some characters have legal definitions. The data subject in early US agency reports is undefined, but the 2004 amended version of the 1978 French law defines the data subject as "an individual to whom the data covered by the processing relate."[27] The "user" in the 2002 EU ePrivacy Directive is "any natural person using a publicly available electronic communications service, for private or business purposes, without necessarily having subscribed to this service."[28] The "consumer" in the California Consumer Privacy Act of 2018 is "a natural person who is a California resident."[29] By focusing on made-up computer characters, I present a careful analysis of how those characters came to be tangled up in cookies, and how they changed over time and space, leading to one of tech policy's worst ruts.

Chapter 2 provides an origin story for the data subject. It begins by describing the burgeoning global computer industry in which the data subject emerged. The chapter emphasizes the state of computing as defined, on one hand, by access limited to a class of engineers, programmers, and operators within a few industries and, on the other, by an established dominance of American companies, namely IBM. Computing technology was used in only a few areas of life, but those areas, like insurance and welfare benefits, were essential. The data subject did not directly interact with this technology and was far removed from the process of creating computer data. So, providing consent for participating in them was not particularly relevant for anything except consent to use the data for additional projects conducted by social scientists later. Focusing on the US, UK, France, and Germany, and touching on Canada and Sweden, the chapter analyzes the first government reports investigating computers and society. It traces the creation and development of the data subject in the US in the late 1960s. It then follows the data subject's transfer across the Atlantic, where other ideas of data protection were being constructed as distinct from European ideas of privacy. There, Germany pushed the data subject to be more powerful and established consent as a form of challenge to even essential data systems like the census, whereas France continued to place limitations on automated data processing that might interfere with a broad sense of human rights. The chapter describes the approval of the 1995 EU Data

Protection Directive and the international embrace of the data subject, with the notable exception of the US. The absence of the data subject and emphasis on other character development stateside would cause significant disputes around the design and governance of cookies.

Like the data subject, the story of the user told in chapter 3 also begins in the mid-1960s, with work on packet-switching networks and innovations for communications technologies. While the data subject is tied to computer databases, the user is connected to networks. The user developed over different computer networks that had different goals, structures, and players, all with privacy implications that play out differently. By tracing network initiatives including Minitel in France, Prestel in England, Bulletin Board Systems in Germany, ARPANET and pre-web online services like America Online in the US, and EUNet across Europe, chapter 3 focuses new attention on whom the networks were designed for, what type of privacy was protected, and through what means. Each transnational network experiment involves domestic goals like updating infrastructure or moving early to dictate standards. Those goals consistently included an effort to update wiretapping laws to protect users from digital interception and efforts to update criminal laws for authorized uses of computers. Through these domestic efforts, the user becomes legally distinct from developer/programmer and hacker. The user as a computing character reveals themself as a prominent figure tied to histories of citizenship and speech in communication networks. A far less passive character than the data subject, the user creates, sends, receives, and manipulates information—and in doing so, consents or authorizes certain types of exposure. When the user is too sophisticated and operates beyond the bounds of authorization, they are a hacker.

The privacy consumer, found in chapter 4, is also a relatively savvy character, capable of consent as choice in a marketplace. Yet, the privacy consumer, more so than the user or the data subject, was intentionally created. The privacy consumer was part of the US Clinton administration's (1993–2001) efforts to grow and commercialize the Swiss-based web. Early signs of the privacy consumer are found in debates over the technical problem of "maintaining state" that were hashed out on web developer and standards email lists. Various actors and motives created the data memory system for the contemporary network (cookies), making references to anonymous users who would choose privacy as consumers. These newly imagined consumers making choices about privacy would become privacy consumers in the hands of Clinton's Federal Trade Commission (FTC). Through the efforts of a growing digital ad industry, advertising was promoted as infrastructure for the new and delicate web, complicating the role consent might play in the system's design. Nonetheless, the Clinton administration sought to capitalize on and establish power over the web through a commercial and consumer framework that would rely on choice. Tech companies from Silicon Valley in California, ad agencies from Silicon Alley in Manhattan, and policymakers on the Hill in Washington, DC, used an old "negative check off system" (opt-out) from the 1970s that had been created to keep regulation out of the direct mail industry. Back then, direct mail was understood to support the US postal service, and the opt-out system had been found satisfactory by earlier government commissions investigating privacy. They promoted a privacy consumer who chooses to participate in targeted advertising that also served as infrastructure for the digital future. The privacy consumer didn't flourish in Europe, however, as digital advertisers working across European countries with different languages, models, and laws faced significant challenges.

Set in the aftermath of the dot-com crash and the coalescing of internet startups, chapter 5 describes domestic policy failures and transatlantic disputes stuck in a tangled mess of computer characters. The cookie notification system now congesting the internet is the manifestation of misunderstanding these characters, legal backgrounds, and technical roots. Cookies are both computer data and network communications, and so they are about privacy *and* data protection. In the US, the two are conflated, and references are frequently made to "data privacy" as a catchall. Further, treatment of "data privacy" often occurs through a "consumer privacy" approach. In Europe, data protection and privacy are not the same and carry different functions in service of two different human rights that have two different histories. The chapter details failed class-action suits filed in the US to combat cookies and notes the lack of US legislative action on the subject. The chapter addresses the EU's ePrivacy Directive, which specifically regulates cookies, and discusses how policymakers sought to update their laws in the face of an increasingly powerful anti-regulatory, pro-innovation American rhetoric. It follows the development of privacy research and privacy-enhancing technologies (PETs) that sought to improve the conditions of consent and helped shape policy debates. The chapter includes a detailed history of the 2009 changes to the ePrivacy Directive, the General Data Protection Regulation, and the EU–US Safe Harbor arrangement. It describes cookies as a consistent point of contention, because the US and EU countries have constructed, transformed, and deployed computer characters inconsistently and without regard for the variable abilities, knowledge, and motives central to consent for each.

The development of these computer characters was and is heavily reliant on historical ideas about the future. Science and technology studies scholars use the theory of "sociotechnical imaginaries" to help understand and articulate how shared narratives of the future reveal and promote certain visions of technology and social order over others.[30] Within these sociotechnical imaginaries, two particular practices can be traced that help reveal the characteristics of computer characters. The first, from technology law, is the "imagined regulatory scene."[31] Every legal principle is developed with a presupposed foundation and set of future scenarios that it anticipates and shapes.[32] Cookies are regulated by laws established for three distinct sociotechnical regulatory scenes. Often research on imaginaries focuses heavily on visions of the future without noting which legacies are included in that vision. The sociotechnical imaginaries through which computer characters emerge carry longstanding ideas about labor, innovation, economics, global competition, and harms. Such legacies are vital to their stories and understanding the consent dilemma as more than a design problem. The second practice within the sociotechnical imaginary, from space anthropologists, is a "projectory." A projectory is "a material instantiation of the 'imaginary,' made concrete and traceable through circulated documents that codify a community's orientation toward future technological states."[33] Looking closely at cookies as projectories in technical communities allows us to see how the computer characters were negotiated as part of disputed futures. Over the story of cookies, the computer characters develop in both the policy setting of the sociotechnical imaginary where they are debated and codified, as well as in the technoscientific realm of the projectory where they are further disputed, negotiated, refined, and distorted. When the terms of sociotechnical imaginaries change, which inevitably involves alterations to the imagined regulatory scene, a material instantiation like the cookie is set aside and new projectories are undertaken to further new visions of the future. Participation in these efforts may involve the creation of new computer characters. Chapters 2 through 4 describe different national imaginaries and projectories and chapter 5 explores how diverse imaginaries and projectories can collide in an international policy environment.

The final chapter of the book, chapter 6, tells the story of the cookie's retirement, driven largely by two major boosts to consent that occurred around 2018 and after, in the wake of significant social unrest and associated "techlash." As cookies are scheduled (by Google) to be phased out in 2024, a number of sociotechnical re-imaginaries vie to "redo" the internet or the web or social media are dueling for the future. Competing re-imaginaries often include versions of the characters described, boosted by legal or technical powers. The first boost came from the EU's General Data Protection Regulation (GDPR), which went into effect in 2018. The GDPR changed the definition of consent, explicitly defining that consent had to be freely given, specific, informed, and unambiguous. Consent had to come before any collection (prior) and manifest through action (affirmative). Because the cookie laws in Europe refer to the definition of consent from European data protection law, cookies suddenly required prior, affirmative consent. Another boost to the data subject came from Illinois's 2008 Biometric Information Privacy Act (BIPA). These new legal requirements were complemented by growing pressure from other tech companies, namely Apple. When Apple announced that its iOS update to the iPhone would require users to select whether each app could share data with other apps before proceeding, Facebook launched a massive ad campaign bashing Apple for ruining the free internet and hurting small business. Facebook was right to be scared—only a tiny portion of iPhone users clicked the "Allow Tracking." In 2021, the French data protection agency, reinforced this technical version of consent, fining Google and Facebook over €200 million for using a consent pop-up that was easier to accept than reject. Such an interpretation of consent can also be found in reports from the Federal Trade Commission seeking to protect privacy consumers from dark patterns, user interface tricks that dupe us into, for example, clicking things we don't want to click. For the broader governance of technology demanded by the societal unease associated with everything from health to equality to democracy, new computer characters should be considered. I suggest two: *non-users* and *members of the public*.

CONSENT CONFUSION

While Zuckerberg was testifying, Facebook was already operating under a US FTC consent decree issued in 2012 after Facebook shared consumer data in ways that the agency found harmful to consumers.[34] The FTC deals with *consumers* within the bounds of consumer protection. Part of the agreement required "affirmative prior consent" to transfer consumers' data to third parties. Surely Facebook violated the agreement when it allowed the data of eighty-seven million Facebook friends to be shared with a third party based on the consent of only 300,000 account holders? When asked before Congress, Zuckerberg simply stated that no, the company did not violate the agreement. The company had complied with the terms of service at the time. And, in fact, the decree arguably did not hold Facebook responsible when consumers forwarded information about others.[35] This would lead one to the odd conclusion that Facebook friends could "consent" to the use of their friends' information. Until 2014, Facebook built this type of consent into the default settings that allowed transfers to app developers. Nonetheless, the FTC investigated and settled with the company for $5 billion in 2019.[36]

Zuckerberg argued that all people should have control of their data. But he also argued that Facebook had not violated its agreement with the FTC. In doing so, he exploited the overlooked differences between computer characters that translate into a profitable type of privacy.

Consent, control, privacy, and choice are all slippery concepts that are often confused, frequently on purpose. Zuckerberg repeatedly answered that Facebook goes above and beyond its privacy policy, developing well-tested and highly effective tools to help people understand what they are sharing and with whom. Congresspeople raised accusatory fingers because of the apparent immorality of collecting, processing, and sharing data without consent, but Zuckerberg easily responded by explaining the many ways that people do consent and even have control. In responding this way, Zuckerberg sidestepped the issue that landed him in the hot seat on the Hill. People have many controls for the information they share on Facebook—but not for the data collected and shared about them. This kind of confusion has led to policy and technology ruts that are mind-numbingly frustrating for scholars but beneficial to tech companies that make money by selling services purportedly better at connecting "data-driven" ads to heavily surveilled eyeballs.

Pulling the concepts apart can be incredibly satisfying. Although consent is attached to privacy, the two are not the same and their relationship looks different in different cultures. In his sprawling but short history of privacy, social historian David Vincent finds privacy in all eras, writing, "There are no beginnings in this history, only threatened endings."[37] Vincent searches for privacy as intimacy, seclusion, and freedom of thought and behavior as he explores changes in demography, family size and makeup, rural and urban populations, housing design, reading habits, diaries, newspapers, gossip, mail, and the telephone. Samuel D. Warren II and Louis Brandeis published perhaps the most famous law review article of all time in 1890, wherein the two argued for the legal acknowledgment of a new right: the right to privacy. Warren and Brandeis's right to privacy was a "right to be left alone," particularly from the press, in one's "private life, habits, acts, and relations."[38] Decades later, in 1960, William Prosser organized cases that sought a right to be left alone, as he understood it, into four distinct privacy torts: intrusion upon seclusion, public disclosure of private facts, false light publicity, and appropriation of name or likeness.[39] Another fifty years later, Daniel Solove created a taxonomy for modern information privacy claims, broken down into four major categories: 1) *information collection* through surveillance or interrogation; 2) *information processing* that results in aggregation of data, identification, insecurity, secondary use, or exclusion; 3) *dissemination of information* that is a breach of promised confidentiality, disclosure of truthful information, exposure of sensitive information, increased accessibility, blackmail, appropriation of one's identity, and distortion of identity through false or misleading personal information; and, 4) *invasions* upon individual's seclusion or decision-making.[40] Consent plays different roles in these privacy categories but can powerfully transform all of them.

Ideas and practices of privacy may differ from place to place, but the concept of privacy endures. In his cross-cultural study of privacy, social psychologist Irwin Altman found the exercise of boundary regulation to be universal. But boundaries weren't the same everywhere, nor were they managed the same way everywhere.[41] We can think of this boundary management as a tension that exists differently, but everywhere. Historian Sarah Igo frames the history of American privacy as a constant and morphing tension between the need to be seen, recognized, and counted, and the desire to maintain boundaries between the social and the individual.[42]

It can also emphasize another tension, one between societal preservation of privacy and a growing American insistence on access, transparency, and knowledge, as journalism scholar Michael Schudson and legal scholar Amy Gajda write.[43] Both tensions are represented in US attitudes toward privacy, but consent to share can dramatically relieve those tensions by limiting one's ability to claim a recognized expectation of privacy.

These tensions exist in other Western privacy claims wherein consent may not change the legal relationship in the same way. European countries package privacy within "personality rights," a path not taken by American jurisprudence. Personality rights are the kinds of protections due to "each person, as a unique and self-determining entity."[44] Compare two cases, one in the US and one in Germany, in which an author of a memoir is sued over details shared about another person within the pages.[45] In *Bonome v. Kaysen*, the author described her unnamed ex-boyfriend as whiny and insensitive about sex, but the court dismissed the lawsuit for public disclosure of private facts. The court found that the author's subject matter was of legitimate public concern, meeting the First Amendment standard that requires only "a sufficient nexus" between private information and public interest. The book was newsworthy.[46] In Germany, a writer shared similar details of her life in a novel, but the German Federal Constitutional Court ordered the book banned. The interference with personality rights established in both the Civil Code (legislation) and the Basic Law for the Federal Republic of Germany (the German constitution) guarantees the protection of dignity by preserving the sphere of intimate life. Under the German constitution, the author's right to artistic expression must be balanced with the plaintiff's right to maintain "subjectivity" and not be "degraded to a mere object."[47]

The idea of control as privacy became popularized in the mid-twentieth century. Over the course of the 1960s and 1970s, data became a possession of the individual—*my* data, *my* cookies.[48] The press portrayed Henrietta Lacks in the 1950s and 1960s as an unwitting source of an important immortal cell line. But the 2010 bestseller *The Immortal Life of Henrietta Lacks* and 2017 HBO movie portrayed her as a woman exploited and rightly owed compensation for the collection and use of her data.[49] Demands for protecting "my data" were the product of social movements, civil rights reform, and market thinking. Articulated most prominently by law and policy scholar Alan Westin in 1967, privacy was "the claim of individuals, groups, or institutions to determine for themselves when, how, and to what extent information about them is communicated to others."[50] Indeed, Altman explained his comparative framework for considering boundary management privacy in terms of control: "I gradually came to speculate, as did Westin on the basis of his analysis of privacy, that the psychological viability of an individual or a group was dependent on the ability to control interactions with others, and that privacy regulation was a culturally pervasive process."[51] Pivotal US privacy reports in the 1970s used the language of control. One states, "Concern about computer-based record keeping usually centers on its implications for personal privacy, and understandably so if privacy is considered to entail control by an individual over the uses made of information about him. In many circumstances in modern life, an individual must either surrender some of that control or forego the services that an organization provides."[52] But US legislators passed only a few narrow laws to address these concerns over data held by certain entities like health institutions, educational facilities, and credit reporting agencies. By default, collecting, processing, or using data does not require the moral or legal magic of consent, nor are there many other limitations on processing.

In Europe, however, legislators incorporated the idea of individual control over computer data into an entirely distinct right of data protection. European data protection is and always has been about power and autonomy. Referred to early on as a "self-help solution" and "subject external control" method, control was not originally tied to consent. It was a tool for possible combination with administrative oversight ("data commissioners"), self-regulation ("voluntary control"), and ex ante licensing and registration schemes.[53] Individual control was one of many methods deployed to address the institutional control and potential control exerted by integrating computing power. Chapter 2 details how and why that changed, but these initial methods of computer governance established a different default in Europe. Although the data protection right in article 8 of the 2000 Charter of Fundamental Rights of the European Union does not use the word "control," and instead focuses on the requirement that data be processed only with consent or another legal basis, data protection laws have a long history of subject controls like the right to know, access, object to, and correct data processing.[54] Within that history are debates about creating a data protection right that is distinct from the right to privacy and uniquely about control.[55] Data protection promotes the value of and interest in control regardless of harms, and under European data protection law, "individual control is a liberal objective which is pursued for liberal aims."[56] The EU Charter frames the right to data protection as a freedom without reference to harms. In so doing, the Charter demonstrates that the right more than safeguards data subjects from abuses. It serves as a source of empowerment in the face of dramatic power asymmetries. Control is the "essence" of data protection.[57]

Consent is also, of course, related to choice. After all, what is consent without choice? Choice is a marker of modern society. Today, "we have no choice but to choose," and selves and lives are considered the direct result of choices made.[58] The "tyranny of choice" not only establishes that everything is a choice but that perfect choices exist and matter.[59] Historian Sophia Rosenfeld explains, choice is:

> Essentially modern life, considered as the hazily defined and largely unanticipated consequence of some blend of Enlightenment notions of the self and freedom, the explosion of consumer culture, the live-and-let-live ethos of the 1960s and, as icing on the cake, the technological revolution associated with the home computer that has multiplied the offerings in every domain from the hundreds or thousands to the millions or billions.[60]

However, choice is limited in important ways. Everyday consumers across societal contexts agree to standardized form contracts (boilerplate), which place them in the "legal universe" created by and for the private organization on the other end of the agreement.[61] Consumers have no opportunity to choose or negotiate the terms of boilerplate. Consumers may be lucky enough to choose one legal universe created by one private entity over another universe, but boilerplates across the board commonly restrict or deny the rights of consumers to demand legal redress.

Economist Richard Thaler and legal scholar Cass Sunstein in their popular book *Nudge*, argue that 1) individuals have limited cognitive capacities, hold malleable preferences distorted by social frames, and operate under many external pressures; and, 2) that private and public institutions should affect behaviors of individuals by manipulating their choices but maintaining choice.[62] Sunstein and Thaler propose carefully crafted "choice architectures" designed to help individuals with limited ability to make choices that maximize their self-interest. But choice architectures built on choice algorithms convert people into *choosers*.[63] Recommender systems, personalization engines, and user analytics "nudge people toward choice-making [serving as] incessant generators of choice."[64] The more time humans spend in electronic contracting environments, the more we are conditioned to choose under the conditions of diminished autonomy in light of the impracticalities of deliberation.[65]

By pulling apart the concepts of control, privacy, and choice, the blurry background actors of consumer, user, and subject come into focus as the legal computer characters of privacy consumer, user, and data subject. Consent as an action word can also be parsed more carefully: the privacy consumer can be seen as a character who consents as choice; the user as a character who consents as an act of authorization; and the data subject as a character who consents as control. Bringing the computer characters to the fore can serve as a stabilizing force, allowing each concept to have its own powerful contours that can help dislodge policy ruts around cookies.

LEGAL CONSTRUCTION OF TECHNOLOGY

Because they have been a curious lightning rod for global privacy debates over so many years, cookies are a valuable lens through which to study a concept like digital consent. To understand the peculiar legacy of cookies, I employ what I have called the legal construction of technology.[66] Inquiries into the legal construction of technology align closely with the social construction of technology developed within science and technology studies (STS), but focus more closely on legal actors, institutions, cultures, and practices. The legal construction of technology rejects the idea that technology has a nature that the law must fight. As legal scholar Julie Cohen writes,

> Technology is not a monolithic, irresistible force ... Information technologies are highly configurable, and their configurability offers multiple points of entry for interested and well-resourced parties to shape their development ... Legal institutions too offer multiple points of entry for economic and political power, and as they are enlisted to help produce the profound economic and sociotechnical transformations that we see all around us, they too are being changed.[67]

Cohen explains that scholars working on information policy have "asked *how law should respond* to changes occurring all around it [but] not asked the broader, reflexive questions about how core legal institutions *are already evolving* in response to the ongoing transformation in our political economy."[68] This echoes STS scholar Sheila Jasanoff's framework for addressing biotechnology policy, which she states is "situated at the intersection of two profoundly destabilizing forces in the way we view the world."[69] Legal construction fits within or alongside the foundational work on coproduction that Jasanoff developed in STS in the 1990s and early 2000s. Coproduction captures how epistemic understandings go hand-in-hand with normative understandings of the world. Scientific beliefs and associated technical artifacts evolve together with the representations, identities, discourses, and institutions that give effect and meaning.[70] "Character" in the title nods to this coproduction, referring to computational representation, people in narratives, and moral nature all at the same time throughout the following pages.

Comparative and historical inquiry through legal construction can attend more fully to these coproductions. Comparison is valuable not as a search for the best decontextualized practice to be inserted into any locale or jurisdiction, but "as a means of investigating the interactions between science and politics, with far-reaching implications for governance in advanced industrial democracies."[71] Even "cultural near neighbors" can present drastically different legal cultures.[72] I focus on the US and EU because they have largely shaped the technical specifications, policy, and experience of cookies. However, other Western countries (notably, Canada) and non-Western countries (namely, China, Japan, and India) have played integral roles in global computing and global internet policy. Why did some legal cultures require consent to compute personal information while others decided that consent was unnecessary for the same practices? How have those normative computing arrangements changed over time? How has it defined our present and entrenched our future?

The role of the individual in exercising their consent, choice, and control in order to protect their own interests and societal goals is central to all technology policies. But politics change. Americans are said to be far more tolerant of commercial invasions of privacy than Europeans, who are far more bothered by the idea of corporate powers. As legal scholar James Whitman writes in his comparison of consumerism versus producerism, "The key identity for Americans is, as so often, the consumer sovereign."[73] Although claims that Europe has finally "discovered the consumer" persist, Whitman insists that paternalistic countries like France and Germany still orient toward producerism, focusing on workers, small business owners, and industry competitors.[74] Whitman finds in his sweeping privacy comparison of the regions that the important difference lies in Americans' commitment to liberty, which demands the government limit informational control over its citizens, versus Europeans' commitment to dignity, which demands a role by the state in preventing informational control.[75]

However, we must be careful with these comparative categories, and especially with how they may manipulate policy discourse. The locus of regulatory innovation and leadership shifted from the US to Europe around 1990.[76] Between 1960 and 1990, the US was "typically one of the first countries to identify new health, safety, and environmental risks and to enact a wide range of stringent and often precautionary standards to prevent or ameliorate them," while European countries were less likely to regulate strictly, if at all, in these areas.[77] Few of the health and environmental regulatory innovations involved individual consent. Many included outright bans on certain types of products and processes. Prior to the 1990s, European countries were the ones complaining about the trade barriers put up by US consumer and environmental standards, but the US has since complained about European regulatory obstacles.[78] In the 1990s, Europeans became "quicker to respond to new risks, more aggressive in pursuing old ones."[79]

Although there is much value in deciphering these differences, cookies and certainly the broader concept of digital consent resist a clean comparison. Relevant technology policy has never been national without reference to our cultural near neighbors. As such, the comparative historical account that follows is messy and transatlantic. By looking to what media scholars have recently called "alternative networks"—meaning networks that predate or coexisted with the commercial internet—and using the legal construction method, I am able to follow twists and turns in the development of the concept of consent to computing and ultimately to arrive at a different perspective on the cookies of today.[80] Looking beyond criticisms of neoliberalism, this perspective reveals cookies as an iterative, international, and disputed legacy in computing. The legal construction of cookies builds on the work of computer historians who have shown how national and global policies have shaped computing. Legal histories of computing raise valuable criticisms of and disputes over technology. Looking to works of computer historians who cover domestic attempts to gain international success (or at least national independence through technological advancements), the legal construction of technology acknowledges these global entanglements and allows for a close study of influence, rejection, adoption, and adaptation.[81]

But this story of cookies also resists the neatness of periodization. Computer historians and technology scholars often take excellent advantage of periodization to organize and present their histories. Global periodization is challenging because different things happen in different orders in different places while also influencing one another. Historians Martin Campbell-Kelly and Daniel Garcia-Swartz overcome this challenge in their sixty-year history of the computer industry by breaking it into fifteen-year segments. Beginning in 1950 and ending in 2010, they move through the initial commercialization of computers to their standardization to the personal computer and finally the commercial internet.[82] My initial attempts to tell the story of cookies using their eras failed, because much of the law that would come to define consent surrounding cookies developed in different legal areas, such as wiretapping and direct mail, with different historical trajectories. For this reason, I have structured this book to follow distinct paired chronicles: computer data paired with the data subject, computer networks with the user, and the commercial web with the privacy consumer. These chronicles share starting points in the mid-1960s and converge just after the dot-com boom. The story of cookies and digital consent continues throughout the first two decades of the twenty-first century, with significant scholarly and political investment in their relation to global technology policy.

In order to pursue the legal construction of technology, a range of materials were collected, organized, synthesized, and analyzed. I draw from multiple legal and regulatory sources, including thousands of documents related to government reports on computers and privacy, as well as hearings, meeting minutes, regulations, and legislation. These documents include surveys of computer companies and public attitudes; testimony from businesses, activists, and academics; analyses conducted by federal agencies; economic and labor studies; and reflections on visits to computer labs. I also consider judicial opinions, party pleadings, and recorded oral arguments in important cases, as well as recent congressional and parliamentary hearings and commentary. I performed archival research on collections, such as the public www-talk mailing list maintained by the World Wide Web Consortium, the Internet Engineering Task Force HTTP working group archives, the Mosaic archive at the University of Illinois, and the Younger Committee Report on Privacy archives at the British National Archives. Many of the developers in these working groups and the policy personnel working on the government committees generously shared their stories with me.

INTRODUCTION

Additionally, I have consulted corporate press releases, personal blog posts, trade publications about data processing and advertising, technical manuals for operating systems, and web analytics—and many other kinds of mundane digital ephemera. Collecting and analyzing these materials from the new theoretical and methodological standpoint of the legal construction of computer characters provides new conceptual tools for resolving consent confusion and even mitigating the consent dilemma.

NEW HISTORY, NEW INTERVENTIONS

Commentary on cookies is abundant—it would be difficult to write a privacy book without mentioning cookies. Law, philosophy, communication, and technology scholars and journalists have all criticized the consent-based privacy paradigm.[83] Information scientists and privacy engineers have shaped how we understand privacy as control, the problems with consent, the limitations of choice, and the use of cookies over the last two decades. No doubt privacy scholars have had enough of cookies. Indeed, much of the field has turned away from individual consent, choice, and/or control, moving with a progressive social wave that focuses on institution-building at various levels of government and structural problems across societal sectors.

Yet, no book has focused solely on digital consent. The material is scattered and the history untouched. The next three chapters of *The Character of Consent* unfold as narratives of each character, followed by a chapter about their problematic convergence and a final chapter describing recent shifts with an eye toward the future. This organization resolves several important inaccuracies and misunderstandings that result from a lack of dedicated historical treatment.

In existing accounts, treatment of consent starts too late. Books that dedicate attention to cookies focus on the development of cookies by Lou Montulli at Netscape and their proliferation across the web.[84] Starting the story of digital consent in the mid-1990s, however, leaves much out. By then, the US and EU had settled into newly reversed regulatory positions and Silicon Valley had been declared the capital for computer talent, culture, and money. Starting here ignores the previous decades, when communications privacy and data protection were developing in Minnesota, Paris, and Hesse (Germany) under very different political and computing contexts. Without a longer timeline, the historiography of digital consent remains driven by technological determinism and American exceptionalism—a lone young programmer at a feisty disruptive startup navigating the unregulated Wild West of cyberspace.

Another fundamental misconception is that notice and consent systems of privacy come from fair information practices (FIPs). This misconception is packaged in a narrative that the FIPs supposedly worked back in the 1970s, but that today's internet simply presents too many terms of service and privacy policies that are too complex and exist among too few alternatives. Commentators frequently critique contemporary laws or proposals as "notice and consent" or "notice and choice" regimes that are based on the 1970s FIPs.[85] Or, they erroneously attribute European origins to notice and consent regimes.[86] They often argue that, given the deluge of new technology, "old-fashioned notice and choice" has become "impossible."[87] As thoroughly detailed in chapter 3, the original FIPs were simply not about consent. In chapter 4, I show how an opt-out system created by a direct mail industry organization in the US during the 1970s had successfully assuaged policymakers' concerns and was later used by the internet advertising industry to allay concerns again when they threatened to regulate web cookies. The explicit politics of the commercialized web comprised consumer rhetoric, practices, and expectations that, in the US, transformed all internet data into consumer data. The internet did not overwhelm a stable digital consent system; rather, digital consent was created for the internet.

This emphasis on *consumer* data has resulted in the misunderstanding that data protection and privacy are interchangeable—they are not. American scholars, policymakers, and commentators often refer to "data privacy."[88] Privacy is sometimes called the value at the core of data protection, while at other times data protection is conceptualized as a subset of privacy.[89] Many European scholars and practitioners hold that privacy is about information on the move, while data protection is about information at rest. This is true, but privacy is also about confidential communication and a right to one's private domain, whereas data protection is about fair processing and control of computer data. With their unique historical trajectories, we can see the value in keeping them distinct and further interrogating their relationship to one another.

The motives of European policymakers are also frequently misconstrued in these historical trajectories. Many criticize European countries for being more interested in regulating computing than innovating. Talking to tech reporter Kara Swisher in 2015, President Obama accused Europeans of economic protectionism, stating, "We have owned the Internet. Our companies have created it, expanded it, perfected it in ways that they can't compete. And oftentimes what is portrayed as high-minded positions on issues sometimes is just designed to carve out some of their commercial interests."[90] While talking to Swisher in 2019, Silicon Valley congressional representative Ro Khanna insisted that the US should not pursue European-style regulation because "Europe is almost irrelevant when it comes to technology . . . I would not listen to anyone in Europe about anything in tech."[91] Despite this anti-innovation reputation, Europeans and Europe have been integral to computer innovations and have developed aggressive policies to capture global markets.

By contrast, moving the historical starting line back a few decades provides a much different history of digital consent than taking up this issue in the 1990s. Digital consent did not become outdated by the internet but was created for the internet. Data protection, consumer protection, and privacy law are meaningfully distinct. The construction of these regimes and characters has been and continues to be part of domestic innovation policies designed to lead the global computer industry.

This history matters. Computer scientists frequently invoke history to justify the need for certain designs or outcomes.[92] Software engineers used the history of the assembly line to argue for the necessity of a linear view process of making software, but there were many other histories to choose from that would have imagined software engineering quite differently. History is used to sell computing "to the government as big machines for scientific research, to business and industry as systems vital to management, or to universities as scientific and technological disciplines."[93] History is also used to sell computer policy.

Representative Greg Walden began the second day of questioning Mark Zuckerberg about Facebook's data scandals by explaining:

> You and your cofounders started a company in your dorm room that's grown to one—be one of the biggest and most successful businesses in the entire world. Through innovation and quintessentially American entrepreneurial spirit, Facebook and the tech companies that have flourished in Silicon Valley join the legacy of great American companies who built our nation, drove our economy forward, and created jobs and opportunity. And you did it all without having

to ask permission from the federal government and with very little regulatory involvement.[94]

Silicon Valley's relationship with and reliance on the US government has no place in Walden's telling.[95] Facebook's success in regions that have regulated personal data for decades disappears from the story.

On the same committee, on the same day, history was used to counter the Ivy League–dropout, American hero, inventor success story to sell a different computer policy. Representative Billy Long inquired about the origins of Facebook as Zuckerberg's next project after one called "Facemash," asking, "You put up pictures of two women, and decide which one was the better—more attractive of the two, is that right?" Zuckerberg responded, "Congressman, that is an accurate description of the prank website that I made when I was a sophomore in college."[96]

The history of digital consent matters to how we ask and answer technology policy questions. By reframing the story, this history reveals new opportunities for intervention. Although some privacy scholars may have moved on, policymakers and much of the public continue to experience and debate privacy in terms of individual consent. As institutional systems and structural constraints are imagined and reimagined in the years to come, we should thoughtfully construct and incorporate individual computer characters into these regimes. We should push back on the manipulation and misrepresentation of these characters. As transatlantic disputes continue, as American Big Tech comes under more regulatory scrutiny, and as cookies are scheduled for retirement, we may understand this historical moment as an opportunity to wade forward more reflectively.

2 COMPUTING THE DATA SUBJECT

Of the three computer characters who consent to cookies, the first origin story to tell is the data subject from data protection law. The data subject—the person referenced by digital personal information and in need of protections from potential wielders of computing power—plays a central yet unacknowledged role in technology policy that has shaped cookies. Although data subjects are now associated with European data protection law, they have their roots in the US. This chapter details how the data subject began as a character imagined to have little inherent power to consent to the creation or use of personal information. Such data was created and used by vast bureaucracies on distant, hulking machines for essential social services like welfare, insurance, and credit. At the earliest stages of data protection, the data subject was at the mercy of a user, who was an evolving combination of technicians, operators, management, and maintenance, and who was, in turn, at the mercy of manufacturers. The longer history told in the following pages starts in the early days of the global computing industry when data subjects did not automatically generate their own data on personal devices. It reveals the marked distinction between data protection, privacy, and consumer protection and illuminates the contours of the data subject. The data subject can be seen as a unique, legally constructed computer character associated with distinct social goals in relation to computers.

The power dynamics of computing—the potential to exert control over individuals by flattening them into data legible by those with computers—became evident in government committee work. The articulation of those power dynamics evolved in committee meetings that reference foreign efforts, technical experts with transnational ties through research outfits or global corporations, and domestic objectives that rarely lost sight of the international players. As these dynamics and rhetoric evolved, computing became a morally distinct act recognized in national laws across Europe. Abroad, American dominance in the computer industry created a foundation for understanding computing as a technology with unique moral arrangements. The data subject was central to these moral arrangements because European countries enacted data protection laws specifically to protect the data subject. Without the same threat of foreign control, the US could not overcome national political challenges around surveillance in the 1970s to pass laws that would establish a legal regime for computing. Since then, the data subject has disappeared from US policy, substituted by the consumer, the patient, the student, the employee, and so forth. Meanwhile, the data subject has gained its own fundamental rights across the Atlantic. As such, this computer character remains an extremely important figure in transatlantic tech disputes.

THE AMERICAN CHALLENGE FOR EUROPEAN COMPUTING

Although many commentators explain the distinction between the US and EU computer policies in terms of Europeans' experiences with Nazi abuse of information and technology during World War II, perhaps the most significant impact that the war had on digital consent was to give the US an immense advantage by leaving it far less decimated. US government and industry could invest heavily in research and development with idiosyncratically strong domestic demand. In the 1950s, European countries, including those best positioned to engage in the computer industry like the UK, Sweden, France, and Germany, had far smaller per capita incomes and populations.[1] With lower wages and stronger unions, Europeans were less inclined to replace labor with automation and were less captivated by office equipment. In contrast, "because [of] the American office's late start compared with Europe, it did not carry the albatross of old-fashioned offices and entrenched archaic working methods . . . But no simple economic explanation can fully account for America's love affair with office machinery."[2] Differences in structural receptivity and cultural enthusiasm for office equipment led to differences in who built, maintained, and operated computers, and how those classes of owners, managers, and workers would contribute to the proliferation of computers. French politician and political writer Jean-Jacques Servan-Schreiber stirred European readers into action when he presented this situation in his international bestseller *Le Défi Américain* (*The American Challenge*, 1967), driving a fresh wave of nationalism and transnational European cooperation. Who would channel the computing power of the nation, he asked, and to whose benefit?

The Western computer industry began with punched-card machines used to tally the census—machines built to count and categorize people. Prior to both world wars, German American inventor Herman Hollerith, inspired by European efforts, developed a mechanical tabulator using punched cards for the 1890 US census. He started a business based on the machine. However, punched-card equipment built by James Powers, a Russian immigrant working as an engineer at the census, processed the 1910 census.[3] He too then started a business based on the machine, one that would have driven Hollerith's company out of the market had Thomas J. Watson Sr., not joined its leadership. Stateside in the 1920s, Hollerith's and Powers's companies competed as IBM and Remington Rand after a series of mergers, against a handful of other competitors. Overseas, the Powers Tabulating Machine Company counted the 1920 British census and expanded throughout Europe as Powers-Samas (combining its distribution through the Accounting and Tabulating Machine Company of Great Britain Limited, with its French sales force called Société Anonyme des Machines à Statistiques). These early machines commercialized counting people, far removed from the equipment or expertise, for states establishing means of allocating government resources.

IBM focused on growth in the 1920s and expanded across Europe, Latin America, and Asia. In Germany, IBM partnered with Willy Heidinger's firm Deutsche Hollerith Maschinen GmbH (Dehomag).[4] Prior to the war, the Nazis were customers of Dehomag and contributed to Heidinger's success across the region. In France, a bank employee bought the patents for a counting and sorting machine invented by a Norwegian engineer named Fredrik Rosing Bull.[5] The bank employee and his punched card supplier started a company to manufacture and market the Bull tabulator machines, Compagnie des Machines Bull, that quickly took on IBM's French subsidiary.[6] In 1910, IBM held 90 percent of the market. Although the company aggressively grew the global market, its share dropped to 80 percent in the

1930s thanks to Powers-Samas and Bull.[7] Great Britain was an exception, though IBM was always the one to beat. The British Tabulating Machine Company (BTM) originated as a licensee of IBM products and later joined other computer manufacturers to hold onto most of the British market for many decades. Although IBM was the de facto supplier of computing technology for companies and government services around the world, these transatlantic arrangements served as the foundation for the international computer industry.

During World War II, Powers-Samas built cipher machines for the British and supported administrative tasks for the Allies, but IBM's tabulating machines, office equipment, and personnel were woven into the fabric of the war. IBM's profits soared. It provided 80 percent of the equipment used for accounting, inventory, and manufacturing. "Every person in uniform required a set of 80-column cards to track their personnel history, training, and assignments from the time they were sworn in until they were discharged," and IBM processed the data from each of them.[8] The war gave IBM engineers experience with advanced electronics and it gave the company opportunities to build new high-profile research partnerships that "inched IBM into the computing world."[9] IBM's national European subsidiaries, on the other hand, essentially became disconnected from the New York headquarters and attempted to maintain operations given their local conditions. In Germany, IBM's subsidiary Dehomag was still led by Heidinger, who actively courted the Nazi party. The outposts sometimes shut down because of equipment confiscation, or worked to repair bicycles instead of tabulation machines while waiting out the fighting.[10]

Organizing and rebuilding after World War II created a great demand for government services and, with it, data processing machines. Although the war shattered scientific infrastructures across Western Europe, massive welfare states that provided expanding government services (like social security, employment, veteran, and health services) and public utilities (like electricity) were established on a foundation of machines. Punched cards were used to input and store data, and, later, to store and input programs to stored-program computers. Punched-card readers used pins, brushes, or light that generated a response (mechanical or electrical) behind the card where holes were present. The automation of punched card tabulation still involved a lot of human labor. First, a data collector like a census enumerator would write information collected from an individual (or, if the cards were being used to program, a programmer would write their program on a special coding sheet). Then, another person organized the information and gave it to a keypunch, who translated the information onto a punched card using a keypunch machine. The keypunch handed the data or program to a computer operator, who would feed the cards into the machine and then deliver the results back to a manager. Until the mid-1980s, perforated paper-based storage for computer processing like punched cards and tape rolls remained in use for data entry and programming even after other forms of storage were available, because they could be changed without accessing a computer.[11]

Beyond punched-card tabulation companies like IBM, several companies were developing machines to perform calculations for their existing markets. Prior to and during the war, "computers" were women performing calculations by hand. Later, machine calculations actually involved many hands as well. Women first programmed these machines, taking days to physically reconfigure and check wires to perform each instruction under absurd conditions and without recognition. After the war, more women joined the growing field, as few men returning home gravitated to the programming jobs that had always been filled by women. These early computers, like the Army-financed Electronic Numerical Integrator and Computer (ENIAC), were "better described as a collection of electronic adding machines and other arithmetic units, which were originally controlled by a web of large electrical cables."[12]

The ENIAC became the basis for a company sold to Remington Rand, which became one of the "Seven Dwarfs" of the computing world (as in "IBM and the Seven Dwarfs," a then-popular but since forgotten nickname). IBM's most celebrated leader, Thomas J. Watson Sr., was previously fired from another, the National Cash Register Company (NCR). Other early competitors were Burroughs, Control Data Corporation (CDC), Honeywell, RCA, and General Electric. They each brought computational equipment and expertise to their established sectors, which included adding machines for accounting, scientific equipment for experiments, cash registers for retail, and typewriters for offices. In the early 1970s, GE sold its computer division to Honeywell, and Remington Rand's parent company bought RCA's computer division.[13] More cannibalization followed, and new competitors that made smaller computers appeared, not just within the US. Jockeying for market shares was becoming an international sport.

After World War II, IBM reorganized and created a wholly owned subsidiary for its foreign operations. Watson stated that IBM's performance abroad was "nothing short of a disgrace" and created "his own internal common market a decade before the Europeans did."[14] IBM World Trade Corporation established wholly owned country-level companies, each called IBM followed by the name of the country (e.g., IBM Spain). Germany, France, and Britain produced the major competitors. As in the US, those competitors came from established markets like punched-card vendors and electrical supply companies, as well as computer startups. Developing a domestic computer industry that could thrive meant fending off IBM or strategically partnering with it or one of the Seven Dwarfs to compete.

In the late 1950s and 1960s, IBM introduced two computers that served such major blows to its competitors that most did not survive: the 1401 and System/360. In 1959, the company debuted the IBM 1401, which found its way to Europe by 1960. The 1401 transitioned IBM and its customers (and competitors) from its punched card business (estimated at 30 percent of its revenue in 1960) to the electronic stored-program computer era.[15] Stored-program computers allowed programmers to load their programs (encoded on punched cards or magnetic tape) without physically changing the machine.[16] With a central processing unit (CPU), a card reader and punch for punched cards, magnetic tape drive, and high-speed printer, the 1401-style offered flexibility and simplified use (though it still had no keyboard or display). IBM designed the 1401 for ordinary businesses at a more affordable price.[17]

System/360 was a knockout blow that landed in 1964. "A short list of the most transformative products of the past century would include it, competing with electricity and the light bulb and with Ford's Model T car," declared historian James Cortada.[18] At the most basic level, System/360 was a family of mainframe computers. Originally the "main frame" physically housed the CPU and memory for these machines, but "mainframe" came to mean a style of computer architecture wherein a powerful centralized machine supports thousands of data processing applications with many input/output devices serving many users. System/360 mainframes had simplified operator consoles for turning the machine on and off and a troubleshooting panel (every machine came with a resident IBM engineer for maintenance and emergencies). It also offered options for a graphical display with keyboard, card processor, paper tape reader, removable disk storage, and a printer. IBM emphasized access and control of data in its marketing materials:

> Revolutionary mass information storage with . . . direct access to billions of characters . . . Enough for your records. *Your* records: Available when and where needed. Available in the computer room or 3000 miles away. System/360 makes them available. Link hundreds of remote terminals to one multiplex channel and then a central installation becomes the hub of a nation-wide network.[19]

Your records, of course, meant your organization's records—records for things like tracking satellites, dispatching electric power, automating oil fields, credit reporting, filling policy status inquiries, updating inventory, and plotting engineering results.

System/360 cost IBM $5 billion, which almost wrecked the company even with its consistently huge revenues and market shares. But it was not something competitors could replicate. System/360 took all of IBM's fragmented and incompatible machine series and streamlined all the company's offerings into a suite of unified, easily expandable systems. Customers could upgrade their equipment without having to rewrite new software. Because the hardware was interoperable, the system was more stable, and software and sales personnel's work was simplified and made more efficient across the six models and the collection of compatible peripherals (e.g., printers, memory storage, communication terminals, input/output devices). By the 1960s, computer work was becoming a profession—a male-dominated and well-paid profession in many places—and each component of the System/360, including software, lent itself to new markets for new talent.[20] It split the computer industry into companies pursuing IBM-compatible products and IBM-incompatible alternatives, as well as small niche computers.

The 1401 and System/360 blows were actually defensive responses to European innovations. "In the mid-1950s, IBM got a wake-up call," explained Chuck Branscomb, who led the 1401 design team. "It was a competitive threat."[21] He was talking about the French Bull's Gamma line of computers. The Gamma 3 was a stored-program computer that was sold as a calculator for banks and accounting firms. Bull had seized on post-war computing advancements and partnered with the Italian business machines company Olivetti to meet the needs of the European business and military markets. With 50 percent of the market wrested from IBM, Bull expanded across Europe and Latin America and exported to the US through Remington Rand. When IBM introduced the 1401 into the French market, Bull scrambled to maintain its competitive edge. By 1963, Bull's failed agreements with Honeywell, the Japanese Nippon Electric Company (NEC), and RCA left it with no choice but to merge with GE. Although French president Charles de Gaulle tried to prevent the merger because Bull was vital to French nuclear developments, Bull-GE formed, only to become Honeywell Bull after GE tired of struggles with IBM and sold its manufacturing plants. After that, American companies held 90 percent of the French computer market.[22] A French computer startup that made scientific computers and a company that sold licensed American equipment shared the rest. After IBM launched System/360 and the US denied export licenses of IBM and CDC computers to the French research agency Commissariat à l'énergie atomique, the national champion Compagnie internationale pour l'informatique (CII) was formed as part of de Gaulle's ambitious Plan Calcul. CII, like Bull, made IBM-incompatible alternatives. Even with extraordinary government investment and involvement, CII lost money and maintained only around 10 percent of the French market to show for its efforts.[23]

The British champion failed too. The electrical manufacturers Ferranti and English Electric produced two university-designed computers. A bakery produced another.[24] IBM and Remington Rand had discontinued their licenses with BTM and Powers-Samas in the 1940s, and the companies merged in 1959 to create International Computers and Tabulators (ICT), which later acquired Ferranti's computer division. After the release of the 1401, the newly established Ministry of Technology (MinTech) encouraged a merger with the two players left: ICT and English Electric Leo Marconi (EELM), both products of earlier mergers. After System/360, Prime Minister Harold Wilson famously called for a new Britain forged in the "white heat" of new technology.[25]

MinTech funded the merger of ICT and EELM to create International Computers Limited (ICL) in 1968. Despite being underfunded during a period of poor economic conditions, ICL had more than half of the British computer market and a "buy British" government policy ensured ICL would continue to get government contracts. ICL's fate changed with continued improvements to the performance of IBM systems and the Thatcher administration's fierce anti-interventionist policy. While European national champion policies are often blamed for lagging European tech markets, historian Mar Hicks argues that British labor policies played a pivotal role in the collapse. They point to exploitative practices over the period that culminated in a displacement campaign designed to replace female technicians with a male professional class, even while the country suffered labor shortages in computing.[26] The strategy failed, and Britain squandered much of its lead in computing by discarding much of the country's talent, knowledge, and skills. The debate over who would maintain and wield computing power—why and under what conditions—was central to British technology policy. As we will see in the next section, whether and how to regulate computers involves questions of who operates or oversees what parts of the process and what type of governance is understood as suitable for those actors, laborers, or professionals.

After the wars, Allied governments placed a ban on German commercial manufacturing of computers that was not lifted until 1955. In addition, occupation, reparations, and division devastated both the Federal Republic of Germany (West Germany) and the German Democratic Republic (East Germany). Over 80 percent of office machinery businesses were in East Germany and thus largely inaccessible, while West Germany had a population four times larger than East Germany and was where many senior engineers and scientists lived.[27] Meanwhile, IBM exported the "IBM family" (as in nuclear family-like workplace arrangements, not the family of computer models) to its foreign subsidiaries as a way of transferring American welfare capitalism. Late to the game of welfare capitalism but like other American companies, IBM offered high salaries, comfortable production facilities, training programs, insurance, recreation, and a culture fixated on exalting its benevolent leader, Thomas J. Watson Sr., as strategies to ward off union organizing.[28] IBM could not dodge the unavoidable labor unions in West Germany, but "Watson's family rhetoric translated his American welfare capitalist approach into the ideas of Catholic social ethics and made them acceptable to IBM's workforce in Germany."[29]

The 1401 became Germany's best-selling computer, and IBM and the Seven Dwarfs held three quarters of the German market. Germans still managed to find success. Between 1935 and 1938, Konrad Zuse, an engineer bored with routine calculations, invented one of the first electromechanical binary computers; by the mid-1950s, he partnered with Remington Rand Switzerland to produce machines and consult.[30] After System/360, Siemens was Germany's leading electronic equipment manufacturer and so the obvious designee of national champion. It received most of the funds the government spent to boost domestic competitiveness and took 13 percent of the market in the late 1960s.[31] With one entrenched international player with a large majority market share and a primary competitor heavily funded by government investment, the situation was not ripe for competition; however, Nixdorf Computer AG came on the scene in the mid-1960s, filling a niche with small business computers for billing. Started by Heinz Nixdorf, who had previously worked for Remington Rand in Frankfurt in the 1950s, the company was so successful internationally that it started making inroads in the US market by the 1970s.

Computers and computer companies came from smaller national markets as well. The Swedish Board for Computing Machinery was established and set to work building a computer in 1948.[32] Although this government board was discontinued, the talent moved on to the private sector, namely to Facit and Saab. In the 1950s and throughout the 1960s, the Swedish state launched the County Computer Project to administer a national register and taxation system. A public debate ensued over which technology to use: the Saab D21 mainframe or the IBM System/360 Model 30. After testing both and auditing the results, "Big Blue [was] beaten."[33] Others found success through unique projects as well.[34] Canadian universities developed technical expertise and manufacturing competence that was underwritten by the military, and Ferranti-Canada built an automatic sorting machine, called an electronic brain, for the Canadian Post Office Department in the mid-1950s.[35] These achievements notwithstanding, attempts to generate a domestic Canadian computer industry fell flat and Canadian subsidiaries were often treated or portrayed as US outposts despite being significant contributors to transnational companies.[36] Finally, Japan was a major player in the history of Western and global computing in the mid-twentieth century, particularly as another non-European foreign competitor. European companies sought to partner with Fujitsu, Hitachi, Toshiba, and NEC to create new competitive designs and components. Instead of supporting a national champion, Japan invested heavily in domestic computer companies by providing loans, subsidies, and tax benefits.

Leading up to World War II, IBM defined the market, but at the time, the market was small: "There were only a few hundred accounts, a few thousand machines, and only a few tens of millions of dollars."[37] As the computer market grew and took shape, it did so differently in different places. In 1955, there were twenty-seven computers in all of Western Europe and

240 in the US.[38] By 1959, there were 610 computers in Western Europe and 3,810 in the US. By 1964, there were 6,000 computers in Western Europe and 19,200 in the US. Despite government sponsorship of foreign champions and an impactful antitrust case at home, IBM held between 60 and 80 percent of computing markets. The Organization for Economic Cooperation and Development (OECD) tasked the Committee for Science Policy with understanding how differences in technology impact economies.[39] It produced the *Gaps in Technology* report in 1966. Indeed, it found major gaps between the US and other OECD countries. The US government spent eight times as much public money on research and development than the European Economic Community (which would become the EU) as a whole, and US industry spent three times as much on R&D as European companies.[40] Ninety percent of all installed electronic computer equipment was American. Europeans needed to invest in research, talent, and products, and they needed to partner up.[41] Published a year later, *The American Challenge* echoed these observations and solutions.

> These figures are important to keep in mind, for electronics is not an ordinary industry: it is the base upon which the next stage of industrial development depends . . . If Europe continues to lag behind in electronics she could cease to be included among the advanced areas of civilization within a single generation. America today still resembles Europe—with a 15-year head start. She belongs to the same industrial society. But in 1980 the United States will have entered another world, and if we fail to catch up, the *Americans will have a monopoly on know-how, science, and power.*[42]

Servan-Schreiber's nationalism was not protectionist, and he was not anti-American. Quite the opposite. He wanted Europe to rise to the challenge through open markets and global investment and, in doing so, to serve national goals. He explained, "Nothing would be more absurd than to treat the American investor as 'guilty,' and to respond by some form of repression . . . The evil is not the capacity of the Americans, but rather the incapacity of the Europeans."[43] Fearing total elimination of the European computer market, the French and German champions joined to create Unidata. The French sabotaged the effort by insisting on another state-arranged merger between CII and Honeywell Bull in 1976, the French state with a 53 percent stake and Honeywell with 47 percent. Germany sank another round of public funding (over $700 million) into Siemens, which was yet to make a profit from computers. By the third round of similarly large public funding in the mid-1970s, Siemens partnered with the Japanese Fujitsu to compete with IBM on large mainframes, and Nixdorf focused on point-of-sale products as it expanded internationally and headed into a new decade of smaller machines.[44] As governments began to address the concerns raised by the increased use of computing, they did so under IBM's reign after years of battle. People who would soon be considered data subjects were far removed from the labor of computing and creation of computer data, which were important parts of postwar rebuilding and international politics.

THE COMPUTER REPORTS

No Western country was unconcerned or unimpressed with computing power. All sought to be *the* place for computing, which meant tactfully working with or challenging IBM. There is no shortage of histories of IBM, which include detailed accounts of its culture, technology, sales force, and involvement in the Holocaust. For the history of digital consent, IBM is relevant because it was a globally dominant American computer company with an imposing and particular computing paradigm that set the backdrop for computer policy for decades. As these policies developed, IBM remained mostly in the background, but American lawyers came to the fore.

European accounts frequently credit the American privacy debate as the initial catalyst for European thinking about data protection, which may strike us as odd today considering the United States' reputation for lackluster data protection laws.[45] In fact, most Western countries produced committee reports in the late 1960s and early 1970s that referenced two American sources: Warren and Brandeis's 1890 "The Right to Privacy" and Alan Westin's 1967 *Privacy and Freedom*, as well as one another's reports. Data protection developments were quite (Western) internationally coproduced. The hailed *HEW Report* was not the first of its kind, and in its preface, it explicitly expresses a concern that had been voiced in Canada's telecommunications report, Sweden's report on automated personal systems, and the National Academy of Sciences' Databanks in a Free Society project—all completed the year prior in 1972.[46] These reports also attempted to address multiple ways individuals were engaging with computers, not only as consumers or creditors or clients, and extended their analysis to both public and private matters. From wherever the sources of unrest and motivation derived, over the course of the 1960s, the entire Western world seemed to have experienced a shifting sense that computers and computing had become "a symbol of the large-scale technology society and its downsides."[47] Westin's definition of privacy as control over one's personal information was slowly wielded in the development of a new area of data protection law, specifically for the data subject. However, this idea of "control" and its relationship to the act of computing and to the data subject are not obvious or static.

The data subject was born in the US, but quickly emigrated to Europe, leaving behind only a few basic principles for information processing. "Data subject" is a term first used by law professor Arthur R. Miller in 1969, during a hearing on privacy, federal questionnaires, and constitutional rights:

> In a computerized environment, the power to control the flow of data about oneself can easily be compromised. On the theoretical level, computer systems and other media that handle personal information are capable of inflicting harm on the data subject in two principle ways: (1) by disseminating evidence of present or past actions or associations to a wider audience than the subject consented to or anticipated (deprivation of access control), and (2) by introducing factual or contextual inaccuracies that create an erroneous impression of the subject's actual conduct or achievements in the minds of those to whom the information is exposed (deprivation accuracy control).[48]

Miller had testified before on similar subjects to a congressional committee that was assessing the proposal for a national data center. Then, the government was using billions of messy punched cards across twenty agencies to administer services. In a move to create "total archives" of groups and modeling for the economy, regional planning, and the military, social scientists proposed the National Data Center.[49] The idea for the center was articulated in the 1965 *Ruggles Report* (*Report of the Committee on the Preservation and Use of Economic Data*), named after committee chair Richard Ruggles, a Yale economist. The report explained, "Without appropriate data, the economist with a computer would be in the same position as a biologist with a powerful microscope but no biological specimens."[50] The report was given to the Bureau of the Budget. The Bureau commissioned Edgar S. Dunn Jr., who was an economics professor and researcher at the Resources for the Future think tank. Dunn was to report on the report. The *Dunn Report* explained that data archives were a mess for everyone involved in research, policy making, or decision-making. It recommended a new system with specific standards that would allow for data reuse.[51] A task force was then created to produce a report on the issue. The *Kaysen Report*, led by economist Carl Kaysen, director of the Institute for Advanced Study, recommended a single facility, integrated data, and expansive access. It also noted that legal measures would need to be taken to protect data from misuse.[52]

By 1965, the House Committee on Government Operations had created a Special Subcommittee on the Invasion of Privacy and the Senate had begun a six-part inquiry into invasions of privacy. Both addressed the National Data Center as part of a collection of concerns about "snooping equipment" used by federal government agencies.[53] In his testimony before Congress, Dunn insisted that he cared deeply about privacy, that the proposal did not involve invasions of privacy because it did not relate to sensitive information, and finally that the possibility of many actors having easy access to individual data existed only in the very distant future, if ever.[54]

The members of both Senate and House committees largely opposed the National Data Center. New Jersey Representative Cornelius E. Gallagher chaired the Special Subcommittee in the House. He feared the "computerized man," who was "stripped of his individuality and privacy . . . his status in society would be measured by the computer, [h]is life, his talent, and his earning capacity would be reduced to a tape with very few alternatives available."[55] Vance Packard, who had brought computing into the public's imagination through his book *The Naked Society* (1964) and would today be called a public intellectual, also testified about the National Data Center. Packard expressed concern that a centralized database about citizens would encourage depersonalization, lead to distrust of and alienation from government, and unfairly keep people from opportunities like job recruitment and government resource allocation.[56] However, Packard wasn't opposed to information collection or the use of computers. He simply wanted safeguards in place. He agreed individuals should be given the right to know— and to correct—what information was available about them. But when pressed, he had no other ideas.[57] On the last day of House testimony in 1966, Paul Baran testified that it didn't matter if they built a centralized data system or not, given that data could be made accessible from anywhere.[58] A Rand Corporation employee who would become known as an internet pioneer, Baran asked the committee to create safeguards to prevent invasions and abuse, but instead the subcommittee tabled the National Data Center proposal and moved on.

When Miller, coiner of the term "data subject," spoke to the Senate the following spring, he agreed with Baran. Miller clarified what he meant by safeguards, warning the committee about consent:

> Excessive reliance should not be placed on what too often is viewed as a universal solvent—the concept of consent. How much attention is the average citizen going to pay to a governmental form requesting consent to record or transmit information? It is extremely unlikely that the full ramifications of the consent will be spelled out in the form; if they were, the document probably would be so complex that the average citizen would find it incomprehensible. Moreover, in many cases the consent will be coerced, not necessarily by threatening a heavy fine or imprisonment, but more subtly by requiring consent as a prerequisite to application for a federal job, contract, or subsidy.[59]

Miller went on to become a legal celebrity, hosting the local television show *Miller's Court* for almost a decade to help the public understand legal issues like government and corporate information access. He published a book in 1971 titled *The Assault on Privacy: Computers, Data Banks, and Dossiers*, and, in a 1972 law review article, he listed government data banks that the press had revealed to the public. He said these data banks were just the tip of the iceberg and warned that President Nixon's welfare reform proposal would give agencies the authority to computerize and exchange data among themselves.[60] Senate judiciary chair Samuel Ervin Jr. pushed Secretary of Health, Education, and Welfare (HEW) Elliot Richardson to consider how agencies should organize people under a computerized system and whether children should be given social security numbers.[61] The welfare reform proposal and Ervin's push led to the HEW Secretary's Advisory Committee on Automated Personal Data Systems, upon which Miller would sit.

The HEW Committee was directed to "analyze the consequences of using computers to keep records about people." In 1973, it produced the *Records, Computers and the Rights of Citizens, Report of the Secretary's Advisory Committee on Automated Personal Data Systems*, widely known in the privacy field as the *HEW Report*.[62] The committee boasted an experienced set of experts, diverse across gender (though not race), academic training, age and background, and geography, whose work continued to define computing in society over the course of their careers. Chaired by computer-security pioneer and founding president of the American Federation of Information Processing Societies Willis H. Ware of the Rand Corporation, it also included additional technical expertise from MIT professor Joseph

Weizenbaum, who had a few years earlier published the early AI program ELIZA and was working on his influential book *Computer Power and Human Reason*, and Guy Dobbs, who was vice president in charge of technical development at Xerox. Carole Parsons served an important leadership role on the HEW Committee.[63] In the 1960s, Parsons had become executive secretary of the National Academy of Sciences' Division of Behavioral Sciences' project to address the known undercount in the census. Using IBM's magnetic typewriters, she drafted and edited a book called *America's Uncounted People* (1972). Through this project, Parsons became familiar with all the government databases in use. She was quickly recruited to the HEW Committee to serve as associate executive director.

In an interesting exchange at the first meeting, Weizenbaum brought up consent in relation to providing social services to "delinquents and predelinquents," as fellow member Patricia Lanphere described them.[64] Lanphere was the Assistant Supervisor of the Bureau of Services to Families and Children in Oklahoma. She explained that when a "client" requested social services, a "service worker" would go to the client's home to help plan their goals. The service worker wrote out the plans on paper and gave a copy to the client. The service worker then handed off the plan to be coded using a form that listed numbers associated with requests under headings like "health," "housing," "individual development," and "education," and another form with action codes to a data processing division. As the plan progressed, it was coded as "initiated" or "referred," and the service worker could get a printout of all the current information and the record of past actions. Dobbs asked, "Does the client after the initial collection of this data—does he get any benefit from the feedback . . . in any direct way?"[65] Lanphere responded that the client only received feedback through the service worker, "He doesn't ever see the computer printout or anything. You know he wouldn't understand them."[66] Dobbs then asked if the clients were told explicitly at the point of collection that their case was going to be managed with a computer. Lanphere said they were, but, "We don't really go into any great lengths to explain the computer because, well, I guess some of the service workers couldn't really explain the computer in great depth . . . [Clients] are really more concerned with what is on this piece of paper: '. . . I need a roof that doesn't leak. I need training for a job. I have a child that's emotionally disturbed and I can't handle him.' You know."[67] This prompted another member to ask why it mattered if the information was on a computer somewhere. Weizenbaum argued that the collection of information involved a general question of informed consent, but if "in fact the collection mechanism and the storage mechanism has certain implications, then the question of informed consent becomes much more difficult than otherwise."[68] Although it wasn't obvious to everyone, Weizenbaum insisted computers changed the nature of consent.

The committee repeatedly presented on and discussed the specialized knowledge and skills held by those who worked with computers, as well as inequalities in the data systems.[69] Parsons noted the large percentage of minority men not counted in the previous decades' censuses, and Miller discussed the prices people might pay for privacy when they could afford it.[70] The group wrestled with what they called the "consent placebo" ("The fact is that if you want some governmental booty you consent," stated Miller), and expressed great skepticism about consent's efficacy.[71]

Testimony from the Society of Certified Data Processors (SCDP) provided an opportunity to further discuss whether there was "some kind of computer related dimension to the automated personal data systems problem which is really quite different than . . . the case if there was no computer."[72] The society representative began by explaining that he tried to keep all his records off computers because they were so error-ridden and messy, but it was impossible. Dobbs asked him to what degree the professional ("whatever that may mean") data processors should be considered ethically or legally responsible. The SCDP was working toward creating standards and, in the meantime, was dealing with complaints. Ware dismissed the whole enterprise:

> This business of certifying programmers or data processors or what-have-you has been long discussed . . . It has never gotten off the ground . . . In the other professions . . . such as engineering, one unfortunately has good theoretical background on which he can act, and the laws of physics will tell you how to design bridges . . . So it never gets down very seriously to a question of subject judgment about whether someone did something wrong or not. In the information-handling process business, there is no theoretical base. It is always subjective judgments and that is part of why I think many people feel that the professionalization and the certification business, either, it is a very long uphill fight, or it is a hopeless fight.[73]

What exactly were these subjective judgments about, and who should make them? When the committee spoke with IBM in June, they were told, "[A] company like IBM does not feel competent to work in a direct fashion in the resolution of the privacy question. That is your job and your assignment in the agencies with whom you are working and to whom your reports and recommendations will be directed, the government policy, and public policy arena."[74] The company felt not only competent but responsible for taking action on computer security issues but privacy was a social issue too large and abstract for it to have a stance on. Indeed "privacy" as a concept was used differently throughout the meetings—sometimes commented upon or critiqued, sometimes not. Richard Gwin, who was working on similar issues in Canada, testified, "My own judgment is that one explanation of this contradiction is that when a number of people use the word 'privacy' and raise complaints about invasions of privacy, they are not talking about privacy at all. They are talking about power."[75] He explained that Canada's first report on computer privacy had come out of a government-sponsored telecommunications conference, as had "a recognition of the virtual impossibility of defining privacy."[76]

As the year came to a close, the team needed to start writing. The committee had received testimony on the Freedom of Information Act (FOIA): "The Freedom of Information Act was designed to turn this whole system upside down so that the question ceased to be 'why disclose,' but rather, 'why not?'" The committee then walked through five exemptions to the government transparency obligations and additional provisions within the promulgating act that prevented disclosure by the agency.[77] Inspired by motives and exemptions in FOIA and principles outlined in the British *Younger Report*, Ware, Parsons, and the executive director came up with five principles that they turned into the Code of Fair Information Practices, a name inspired by the Code of Fair Labor Practices.[78] The five principles stated in the *HEW Report* FIPs were:

- There must be no personal data record keeping systems whose very existence is secret.
- There must be a way for an individual to find out what information about him is in a record and how it is used.
- There must be a way for an individual to prevent information about him that was obtained for one purpose from being used or made available for other purposes without his consent.
- There must be a way for an individual to correct or amend a record of identifiable information about him.

- Any organization creating, maintaining, using, or disseminating records of identifiable personal data must assure the reliability of the data for their intended use and must take precautions to prevent misuse of the data.[79]

These principles were close to those drafted by a subcommittee early in the process. That group had created a "thematic outline" that listed functional characteristics to define automated personal data systems, which included collection, indexing, assembly, and storage of data in machine readable form, as well as electronically controlled manipulation, retrieval, dissemination, or other use of data.[80] The final *HEW Report* uses "computer" or "computer-based" interchangeably with "automated personal data systems" and defines them as "a collection of records containing personal data that can be associated with identifiable individuals, and that are stored, in whole or in part, in computer-accessible files."[81] Watson at IBM adamantly rejected the word "computer" for his 1940s and 1950s machines, insisting instead on "calculators" so as not to suggest workers would be replaced, but "automated" or "automatic data processing" had come to be more commonly used to describe machines that managed data.[82] The report expressed the fear that computers would encourage what is today called "technological solutionism":

> One fact seems clear, however; systems with preemptive potential are typically developed by organizations, and groups of organizations, who see them primarily as attractive technological solutions to complex social problems. The individuals that the systems ultimately affect, the people about whom notations are made, the people who are being labelled and numbered, have, by comparison, a very weak role in determining whether many of these systems should exist, what data they should contain, and how they should be used.[83]

The committee's skepticism toward the personnel who developed these systems made its way into the final report. In a section titled "Technicians as Record Keepers," the committee wrote:

> The presence of a specialized group of data-processing professionals in an organization can create a constituency within the organization whose interests are served by any increase in data use, without much regard for the intrinsic value of the increased use. Another, potentially more serious, consequence of putting record keeping in the hands of a new class of data-processing specialists is that questions of record-keeping practice which involve issues of social policy are sometimes treated as if they were nothing more than questions of efficient technique.[84]

Given the lack of control perpetuated by "the impersonal data system, and faceless users of the information it contains," the organizations that were integrating computing, along with their technicians and managers—not the data subjects—needed education and guidelines in order to gain the trust of the public.

In the draft memo, people are simply referred to as individuals, but throughout the final *HEW Report*, individuals are called "data subjects."[85] Consent by data subjects was not part of the original five principles, unless data was going to be transferred by the organization maintaining the data or used beyond the original purpose. A short clarifying conversation about the wording and meaning of this principle occurred, wherein Weizenbaum emphasized and wanted the language to show clearly that "it is a right to limit something, not the right to extend something."[86] The report, which did not distinguish between public and private data processing, assumed that the maintenance and use of personal data by the initial collector was morally and socially justified without need for consent. The foreword, drafted by Richardson for the MIT Press publication of the report, states, "Most of the activities that generate records about individuals are presumably desired forms of participation in our society. In some instances, they may even be the determinants of survival ... He or she is hardly in a position to refuse to provide information requested by the organization whose assistance is being sought."[87] The scene does not contain consent. And the first sentence of the recommendations in the final report argues its administrative restraints be legally applied to computers: "We recommend the enactment of legislation establishing a Code of Fair Information practice for all automated personal data systems."[88]

The American FIPs differed from the British principles, according to Ware. The British principles were contained in the *Younger Report*. As in the US, the British public shared growing concerns over data that was collected, stored, and connected by government entities, but the Younger Committee was directed only to assess private use of computers.[89] In Britain, the tone darkened between the 1940s and 1960s as computers were integrated into more private sector activities. In a 1965 article titled "Robots Will Set Us Free to Serve," the *London Times* warned that the computer would replace the bomb as the focal point of the public's anxiety: "Little is known and much is rumored."[90] In 1947, the paper had cheerily presented the computer: "All who are prone to come to grief over the simplest sum will be cheered to learn that there has just been invented an electronic computer which can work about 100,000 times faster than man's mind." Even then, though, the paper had concluded, "Some might feel the existence of such machines as a derogation of human dignity. There would seem to be no alternative."[91] Computers continued to make their way into different contexts, where they were met not with coherent or cohesive objections but growing unease.[92] Mixed into a larger debate about government abuses and press invasions, computers were added to a list of concerns over technologies that threatened to leave no space for seclusion, with numerous and continual references to the Americans whom Warren and Brandeis had argued had a right to be left alone.[93] But, computers also integrated control of personal information as part of privacy into the debate.[94] The two proposed UK bills on the subject provided controls through the registration of computers.[95] At the end of the 1960s, more attempts were being made to pass privacy laws, but parliamentary leaders determined the issues needed further study. Kenneth Younger was appointed chair of the effort "to consider whether legislation is needed to give further protection to the individual citizen and to commercial and industrial interests against intrusions into privacy by private persons or organisations, or by companies, and to make recommendations."[96]

The Younger Committee was to assess privacy across many social contexts. They saw the first aspect of privacy to be "freedom from intrusion upon oneself, one's home, family and relationships," but they lacked a proper term for what they considered the second aspect: "the right to determine for oneself how and to what extent information about oneself is communicated to others."[97] Other European countries had begun to build data protection regimes to address the latter, but the *Younger Report* found that, with minor tweaks here and there, existing laws would suffice. It found that with some adjustments, British law could protect fourteen areas of privacy intrusions it had investigated, which included the press, credit reporting, employment, education, nosy neighbors, advertising, and computers. The committee focused on and recommended procedure over substance. For computers, it recommended a voluntary code of conduct for handling computerized personal information and a permanent governmental body to continue to report on developments.[98]

The *Younger Report* was published in 1972, just ahead of the *HEW Report*. Ware read the ten safeguards documented in the British report as similar but "much less specific and not as crisply stated as the provisions in the Fair Code."[99] The *Younger Report* also referred to its safeguards as "principles," rather than "code," which Ware found to be a meaningful distinction.[100] Ware mentioned that the Younger Committee additionally adopted the British Computer Society's (BCS) code of conduct. More than that, the Younger Committee decided on a policy of professionalization, relying heavily on the BCS's vision and governance structure.

The BCS was founded in 1957 to represent "both computer manufacturers and computer users," and saw as its duty to "advise its members as to a suitable code of ethics and professional standards."[101] An important legitimizing organization, it became a professional organization in the late 1960s, requiring professional qualifications for entry of its 16,000 members, about a third of those in the field at the time.[102] In 1971, the BCS approved its code of conduct, a year after its official coat of arms (see figure 2.1). The code of conduct included five principles: to behave always with integrity, complete discretion, strict impartiality, and full responsibility, and without seeking personal advantage to the detriment of the Society.[103] Commentary written by then-BCS President Alex d'Agapeyeff, a Russian-born English cryptographer who cofounded and led the successful Computer Analysts and Programmers (CAP) company, supplemented the code.[104] In it, d'Agapeyeff wrote, "The grounds [for professionalism in computing] are not primarily related to the interests of practitioners. On the contrary, they stem from the requirements that automation should be the effective servant of the community."[105] BCS members testified on the state and future of computing, explaining that there was a movement toward networked systems tied to terminals and that "in such circumstances the use of unqualified people would be disastrous."[106] Professionalization requiring a high standard of skill and knowledge, a confidential relationship with clients, standards of practice to promote public trust, and an enforceable ethical code could avert disaster.[107]

Qualified computer professionals would determine the processing of data, but they needed clear principles about when they could properly transfer data. The BCS testimony emphasized that transferring data could have significant benefits for the individual, with troubling exceptions only in very rare circumstances.[108] Particularly concerned about liability for its members and its industry, the BCS envisioned and designed a Data Bank Licensing Authority for the committee that included a code of

FIGURE 2.1
The British Computer Society coat of arms comprises two core memory store matrices, a lion's face crest that depicts vigilance over the integrity of the society and members, and a key in the lion's mouth that denotes data retrieval and unlocking knowledge, as well as a recognition of concerns the Society holds over problems of computer privacy and security. Source: BCS, *Privacy and the Computer—Steps to Practicality: A Review of Recent Work Carried Out by the Privacy and Public Welfare Committee of the British Computer Society*, London, 1972 (details of symbolic meaning on crest is included in final unnumbered pages of the report).

conduct, constitution, and operations. D'Agapeyeff wrote, "The law used to be interpreted in a clear-cut way. The duty lay in serving the person paying the bills and seldom to anyone else. However the law is changing in sympathy with the public mood which is toward making the professionals more widely responsible."[109] The 1971 BCS Code of Conduct Notes for Guidance stated,

> Members should have regard to the effect of computer based systems, in so far as these are known to them, on the basic human rights of individuals whether within the organisation, its customer or supply or among the general public. Where it is possible that decisions can be made within a computer based system which could adversely affect the social security, work or career of an individual, this decision should in each case be confirmed by a responsible executive who will remain accountable for that decision.[110]

In the UK, professionalization and corporate accountability, not engagement with or input from a data subject, would protect human rights and the general public.

US-based organizations like the Association for Computing Machinery (ACM) were also revamping their codes of conduct and ethics to account for the rights of the public.[111] When updating its 1966 Guidelines for Professional Conduct in Information Processing, the ACM added "Ethical Considerations" to its 1972 Bylaws that stated, "An ACM member, whenever dealing with data concerning individuals, shall always consider the principle of the individual's privacy and seek the following: to minimize the data collected, to limit authorized access to the data, to provide proper security for the data, to determine the required retention period of the data, [and] to ensure proper disposal of the data."[112] The addition made no mention of consent by individuals but charged members to be mindful of their existence and welfare. The *Younger Report*'s ten safeguards for "handling personal information by computers" asked that "appropriate authorization" be given when information was used for purposes other than the one for which it was maintained. Otherwise, it too made no mention of consent.[113] The BCS had promoted an idea of data ownership and argued that transparency and correction were important. It proposed providing a printout made available to the individual upon request. In fact, the Younger Committee debated a "right of printout," but was convinced by ICL representatives that it was possible yet impractical and expensive, given that many installations did not have printers.[114]

Because the Younger Committee investigated "computers" as one of its fourteen areas, in contrast to the HEW Committee's review of computerization across several areas, it engaged explicitly with the computer industry. The committee took tours of IBM UK facilities and, with the help of IBM, it set up a working group to understand the technical aspects of the industry in the country, which had 6,075 computers as of 1971.[115] In its most important distinction from the *HEW Report*, the Younger Committee did not recommend legislation. In fact, it rejected two specific bills that provided detailed controls and oversight.[116] The committee determined that legislation was premature based largely on an absence of identifiable harm. Instead, it encouraged the computer industry to adopt its recommended safeguards, confident that organizations like the BCS would enforce the provisions among their growing ranks. In contrast to the HEW Committee, which had expressed skepticism and uncertainty toward the representative from the Certified Data Processors, the *Younger Report* relied on further professionalization until the need for law was obvious.[117]

At the end of 1975, the government responded to the *Younger Report* with the *Computers and Privacy White Paper*, which promised legislation because "the time has come when those who use computers to handle personal information, however responsible they are, can longer remain the sole judges of whether their own systems adequately safeguard personal privacy."[118] By this point, the UK had become known to others as a data haven, prompting some of the first debates about export controls and concern that other countries might block data to the UK.[119] The Data Protection Committee, called the Lindop Committee after its chair Sir Norman Lindop, took up public and private sector computer systems in the summer of 1976. One key member was Paul Sieghart, a barrister who had written much of the White Paper. As Sieghart had done in his text, the committee separated privacy law from "data protection," a term that, as I discuss in the next section, had become popular after the German state of Hesse passed the first laws in 1970 as Datenschutzgesetz. The Lindop Report referred to data subjects throughout, recommending they be told details about how and why their personal data was being automatically handled. It recommended mandatory registration for all data users. Although the Lindop Report called for decisive action when it was published in 1978, former committee members had to lobby politicians wary of creating additional bureaucracy for years before legislation finally passed in 1984.[120]

In Sweden, computers were introduced into public administration relatively early, and records of citizens were numerous, detailed, and organized by a 10-digit identity number. With a rich history of public access tied neatly to the freedom of the press and therefore to individual expression, the Swedish government began facing concerns in the late 1960s over the increased availability of computerization provided for advertisers, insurance providers, other individuals, foreign countries, and so on. Sweden had already integrated centralized automated systems that other Western countries were debating and rejecting. Swedish historians have categorized computer policy over this period as "rationalizing" versus "critical." Rationalizing policy efforts emphasized computers as positive and obvious paths toward social goods like efficiency and accuracy, while critical policy efforts highlighted the problems and risks associated with computers.[121] Early initiatives in the 1960s rationalized a vision of new technology that would decrease public spending and increase productivity, while later objectives were critical of technology.[122] Two bills that took data protection more seriously passed in 1969 and established the Parliamentary Commission on Publicity and Secrecy Legislation (OSK) just in time to observe and address the public outcry over the 1970 census. The census was the first in Sweden to be computerized, but not the first census (two had been conducted in the 1960s). The event served as a legislative catalyst.

The OSK was instructed to provide a comprehensive overhaul of access and secrecy regulations, paying particular care to computer technology, and to determine whether Sweden needed a specific data regulation law.[123] The task was twofold: "In essence, the OSK mandate was to accelerate and apply the brakes at the same time. On the one hand, the committee was to propose how the principle of public access to official documents could be extended to apply also to computer media; on the other, it had to investigate whether public control should be limited so that people's privacy would not be encroached in the data processing of personal data."[124] The census controversy saved the OSK from having to survey people on their thoughts about privacy. The OSK became high-profile but not high-ranking, as it was made up of lower-level non-political administrators and experts on the ground. Their initial Data and Privacy Report (sometimes translated as Computers and Privacy) resolved that computer files would get the same treatment as other records under the public access law, but that a new law was needed for handling those records.[125] The draft recommended prohibiting the creation of new computer files that contained personal data without permission from a new authority, the Swedish Data Inspection Board (DIB). The DIB would register, inspect, and license "personal registers." A licensing system was necessary, the OSK argued, to determine the state of data processing.

France borrowed from Sweden but was more ambitious. Like many other parts of French computing history, the origins of data protection in France begin with the 1966 Plan Calcul, which set out to do more than boost domestic manufacturing of computers. It was a plan for independence. While the US was spending its resources sending men to the moon, working to develop intercontinental ballistic missiles, and miniaturizing electronics, France was building an infrastructure to integrate computing into every part of everyday life.[126] A research and training component was necessary, so the government created the Institut de Recherche en Informatique et en Automatique (IRIA, French Institute for Research in Computer Science and Automation). Up the hall from IRIA, civil servants were trained in information technology at the Centre d'Etudes Pratiques d'Informatique et d'Automatique (CEPIA, Center for Practical Studies in Informatics and Automation). At CEPIA, Marie Georges, educated as an economist and statistician, and Philippe Lemoine, with degrees in political science and law, sought administrators, managers, and union leaders to develop and promote computer training. They also recruited technical people to teach courses on computers for management. One such leader was Louis Joinet, who founded a union of judges and worked with CEPIA to train those within the Ministry of Justice on computers. One such instructor was the lead engineer on a project called Système Automatisé pour les Fichiers Administratifs et Répertoires des Individus (SAFARI).[127] SAFARI was a system being built by the Institut national de la statistique (INS) that resembled the vision for a National Data Center in the US. It planned to use the country's social security numbers as a unique individual identifier that would allow data matching across public administrations. When Joinet, Lemoine, and Georges found out about the program, they contacted a friend at *Le Monde* and the story ran the next day with the headline "Safari, or the Hunt for Frenchmen."[128] More news coverage followed, inciting government committee work, if not public outrage.

In response, the Ministry of Justice established the Commission on Data Processing and Freedom, led by Bernard Tricot and so referred to as the

Tricot Commission.[129] The high-powered team included Lemoine and Joinet, and was to create "measures to ensure that the development of data processing in the semi-public and private sectors will take place in the context of respect for private life, individual liberties and public liberties."[130] Two of the Tricot Commission members published a book in 1976, *Le Secret Des Fichiers*, based on surveys regarding computer use and public concern.[131] They found that, even after the SAFARI ordeal, the French public was not terribly concerned about computers and potential abuses of privacy. Additionally, the authors concluded the danger was in the information itself, not the computers. Nonetheless, the commission found that legislative action was necessary to prevent the abuse of computers, and that "the major threats seem to be an increase of social control and a worsening of already unequal relationships within society."[132] It recommended a Swedish-style oversight unit and structure that would ensure computer projects were not popping up with no one noticing or remembering.[133] Most data processing would be tied to the purpose declared to the oversight agency, and going beyond that would require new permission from the administration.[134]

The following year, the Ministry of Industry appointed Lemoine to manage the Computerization of Society project. In this role, he helped draft the Nora-Minc Report, which was produced at the behest of President Giscard d'Estaing, who asked Simon Nora to survey the effects of computers. With fellow inspector of finance Alain Minc, the two presented a French society reordered around computers. It was a complete plan for French progress through technological investment. The report was printed as a book and became a bestseller in France, as well as being translated and sold in the US.[135] Over the same period, Lemoine worked closely with trade unions, and in 1977 he coauthored, under a pen name, *Les dégâts du progrès* (*The Damage of Progress*).[136] The book helped establish works council arrangements, which require consultation between management and workers on technology projects and integration of automation and computers.

Back in the US, the new HEW Secretary didn't appear to be interested in data protection. Parsons planned to move over to the Justice Department, where Elliot Richardson was now Attorney General, with intentions of working on privacy policies, but Richardson then abruptly resigned instead of firing the special prosecutor in the Nixon Watergate scandal.[137] After the disclosure of President Nixon's immense surveillance measures and under severe pressure, Nixon set up a commission on privacy, but no one organizing it knew what to do so they again called Parsons. Senator Sam Ervin had proposed a radical privacy bill, and Congressmen Ed Koch and Barry Goldwater Jr. a tamer one. Parsons spent her time negotiating with the parties on behalf of the administration. She went to France and Sweden to exchange ideas about legislation. When the Privacy Act of 1974 passed, it directed the creation of a commission to produce another report "of the data banks, automatic data processing programs, and information systems of governmental, regional, and private organizations, in order to determine the standards and procedures in force for the protection of personal information."[138] Parsons served as executive director, and Ware was appointed as a commissioner on the Privacy Protection Study Commission (PPSC). David Linowes, partner of an accounting firm and economics professor, was appointed chair and was most interested in addressing the hot topic of junk mail. Koch and Goldwater sat on the commission as well, along with William Bailey, president of the insurance company Aetna Life & Casualty, and Minnesota state senator Robert Tennessen. Minnesota had an incredibly vibrant computing community, with a number of broad educational efforts and company divisions. Under Tennessen's leadership, the state passed the Minnesota Fair Information Practices Act (including what is called a "Tennessen warning notice"), making him a valuable resource for federal policymakers.

Published in 1977, the *PPSC Report* established what would happen in the coming decades, far more so than the *HEW Report*. Given the post-Nixon political climate, creating another government agency or providing the government with power to interfere with private organizations was a nonstarter. The hearings were organized around topics: mailing lists, credit cards, tax returns, lending institutions, insurance, medical records, credit reporting, educational institutions, employment records, research, social services, and investigative reporting agencies. The report created a conceptual framework for minimizing intrusiveness, maximizing fairness, and creating enforceable confidentiality. The framework relied "at its base on strengthening the social relationships between individuals and record-keeping organizations by articulating enforceable rights and responsibilities."[139]

The PPSC staff members worked on word processors in the office, but, in the mid-1970s, computers still needed their own rooms, and the members of the PPSC (with the exception of Ware, an electrical engineer who had already made pioneering contributions to computing) were far removed from the machines themselves.[140] The study wasn't really about machines, nor was the framework. It was about relationships. As the report itself puts it, "The Commission finds that as records continue to supplant face-to-face encounters in our society, there has been no compensating tendency to give the individual the kind of control over the collection, use, and disclosure of information about him that his face-to-face encounters normally entail."[141] Koch wanted to limit or prohibit certain types of collection and the PPSC recommended that collection of specific types of information be prohibited in certain contexts (e.g., employers should not collect for use in hiring and promotion). Unlike the HEW Committee, the PPSC feared that as the value of information became more obvious, "public and private organizations may increasingly argue that the impact of allowing individuals to participate in deciding what organizations do with personal information are greater than society can bear."[142] The PPSC thus set out to protect that individual participation. Between the *HEW Report* and PPSC, the data subject as a vulnerable and compromised computer character disappears and sector-specific empowerment of the individual replaces them.

Issues would have to be addressed as they arose within existing administrative structures but would need to be tweaked to promote individual participation in a given context. The PPSC made recommendations for some areas and cautioned against creating government mechanisms to address complaints; instead, it preferred that "such concerns [be] addressed to the greatest extent by enabling the individual to balance what are essential competing interests within his own scheme of values."[143] The *PPSC Report* largely moved away from the data subject, invoking the character only in a chapter on research and statistical studies. The introduction refers to the "individual," and each chapter diverts this figure into their relationships to different data holders. The individual is a "patient" in relation to medical care providers, a "student" in relation to an educational institution, a "creditee" in relation to a credit reporting agency, an "employee" in relation to an employer, a "policyholder" in relation to an insurer, and so on. The PPSC made 162 specific recommendations within each of these contexts, assessing voluntary compliance, statutory creation, or ongoing governmental study. The commission recommended changes to existing law in most areas.

THE COMPUTER LAWS

Legislators in Western countries passed the first laws to address the broad social issues of computing in the 1970s. The laws were new, distinct from the existing privacy torts and communication protections. Although they were sometimes motivated by scandal, many of these laws were rationalizing legislation, so emphasized the promises of computing and sought to enable seamless introduction to make processes more efficient. Even critical laws that established extensive safeguards around automated decision-making were part of integrating computing into society in a way that protected and promoted domestic interests. Different regulatory tools were created to address the power dynamics outlined in the reports.[144] Consent rarely made sense given the great distance between data subjects and the terminal, the machinery, and the memory.

While most countries were organizing committees to write reports, West Germany already had laws on the books. Germany was the first country to apply consent to computing in a sophisticated and purposeful manner, but it did not do so initially. Depending on how one defines such things, the world's first data protection law was adopted in 1970 at the state level in Hesse, where it was unclear who was to be in charge of electronic data banks. The Hessen Datenschutzgesetz (Data Protection Act) applied only to official files of the government of Hesse and established a Data Protection Commissioner. The commissioner made sure government data was not shared or stored in a way that could be accessed or destroyed by an unauthorized person. They also assessed the impact of automated data processing on state operations and decision-making. Written by Spiros Simitis, named Hesse's Chief Data Protection Commission in 1975, the law prohibited "persons responsible" for the processing of personal data from disclosing the data unless another legal provision allowed or someone with appropriate authority "consented" to the sharing. Such authority would derive from managerial status in an organization, not from the *Betroffener*, translated at the time as "person concerned."

The *HEW Report* commented that "the Data Protection Act of Hesse seems designed more to protect the integrity of State data and State government than to protect the people of the State."[145] But Simitis saw it differently, "The [Hesse] data processing regulation ... had been justified by its reference to the growing opportunity to manipulate an individual's behavior through the increasingly sophisticated processing of personal data."[146] Whether rationalizing to protect the state or critical to protect the individual, other German states passed similar laws. Drafts of federal legislation concerning the misuse of computer data bank information in the private sector had already been presented in 1970, but the cost of compliance was contentious enough to delay the issue. When legislation finally passed in 1977, it covered both public and private sectors and had different provisions for government agencies, government-controlled businesses, and private parties processing data for internal use or on behalf of others.[147] Legal scholar Paul Schwartz joked, "Indeed, this statute makes more than its fair contribution to upholding German law's reputation for complexity."[148]

Automated data processing storage required express authorization from a statute or the written consent of the data subject themself. The definition of personal data in the statute referenced the *Betroffener*, still translated as "person concerned" in 1977: "For the purposes of this law, personal data are individual details about personal or factual circumstances of a specific or identifiable natural person (person concerned)."[149] Section 4 lists the data subject's additional rights, including transparency, correction, and objection. The German law transformed the data subject by empowering them to actively participate or prevent participation in computer-based systems. No longer forced to accept automation, the data subject was envisioned as a political actor who could prevent the use of computing. Consent was a central and stable form of legitimizing data processing in the first version of the Federal Data Protection Act in 1977 and in many subsequent iterations.

Computing prompted a new, fruitful wave of thinking about individual rights and Germany's unique contribution to postwar Europe. In 1983, the German Federal Constitutional Court articulated what is now the foundation for data protection in the country: the right to free development of personality and informational self-determination.[150] These remain foreign concepts for many in the US, who may find a combination of privacy and defamation the closest equivalents. European personality rights protect one's identity and dignity, and the court articulated information self-determination such that personal information is tightly tied to one's ability to freely develop identity and maintain dignity in the processing of their data. After the law authorizing the census passed quietly in 1982, significant public unrest grew into full-blown boycotts in January and February of 1983. Signs in windows around the country read, "Beggars, peddlers and census numerators forbidden" and protest banners declared, "Only sheep let themselves be counted."[151] Prior to the census controversy, privacy in Germany had been the domain of bureaucrats and academics, but there was a growing discomfort with government surveillance in the form of computer data banks used to calculate and match records. The census became the catalyst for a swift and deep plunge into the privacy issues of the day.

Two weeks before the census was to begin, the Federal Constitutional Court issued a temporary injunction preventing it, explaining its departure from a right to privacy based on protecting a private sphere where one could be left alone. The Court instead recognized the social nature of information and called for organizational and procedural regulations, but it did not relegate the individual—the data subject—to the background, and instead encouraged them to take part.[152] Data protection was declared strongly entrenched in the basic right of article 2(1), free development of personality, in conjunction with article 1(1), human dignity, of the Basic Law. In this respect, German constitutional rights fiercely protected the power of individuals to decide for themselves the terms of disclosure and use of personal data. Schwartz explains that the data subject was granted power through these rights, but the state was also given responsibilities: "Rather than giving exclusive control over or a property interest to the data subject, the right of informational self-determination compels the State to organize data processing so that personal autonomy will be respected."[153] With a recent history of repressive government actions, Germany, like France, was overtly occupied with issues of power, and legal justifications had to be articulated in the rule of law. As such, German data protection was less reliant on institutions to determine what constituted a morally justified computational action, but it also sought to limit abuse of informational relationships by appointing a commissioner.[154]

Working in parallel, and also within its own domestic drama, Sweden passed its first national data protection law in 1973, following the design set forth by the OSK. The proposed bill emphasized it was only the first step in a continual process to be strengthened as technology developed. Sweden set up a Data Inspection Board (DIB) that issued permits for the creation and maintenance of almost all computerized personal information record-keeping systems in the public and private sector and would also be responsible for the ongoing process of making regulatory and legislative proposals. It did not apply to manual paper files. With a population of only about eight million people and a culture committed to consensus and a high standard of living, Sweden had a relatively simple governmental structure that was conducive to automation. It was "a paradise for registers."[155] Indeed, the DIB director was not hostile toward computers, but explained, "We need traffic rules."[156] The DIB would determine whether a computer system threatened "undue encroachment on the privacy of the individuals."[157] The regime made little mention of the individual. On the rare occasion the DIB would grant a license to link data in different registers (e.g., a union database and an insurance database), written consent from each individual would have to be received.[158] The licensing approach was influential, but ten years later the DIB abandoned the scheme. Many politicians opposed this discontinuation in light of rising public concern over computers and would have given the DIB more resources to be effective instead. Proponents for dismantling the licensing system explained, "It is not possible, we found out, to grant a license to every file."[159] They argued people were familiar with computers at that point in the 1980s and, with consent, found them "quite harmless."[160]

In the US, the Watergate scandal and ensuing impeachment engrossed the country. Nixon's vice president, Gerald Ford, who had sat as chair of the Domestic Council Committee on the Right of Privacy, signed the Privacy Act of 1974 into law after Nixon resigned. The US Privacy Act aimed to protect "individuals to whom the record pertains" in relation to their government. The law responded to the *HEW Report* recommendations (though it was an incredibly narrow implementation of the report's suggested legislative action), as well as public concerns and two congressional hearings. It applied specifically to information held by government agencies. It required agency data banks to provide notice of the nature and uses of personal information and to receive consent from the data subject prior to disseminating the information to a third party. Agencies were not to collect, solicit, or maintain personal information unless it was relevant and necessary for statutory purposes, but consent, which had to be written, was only required for further disclosure.

The first section, "Records Maintained on Individuals," defines an individual as "a citizen of the United States or an alien lawfully admitted for permanent residence."[161] This individual differs from "any person" in the Freedom of Information Act, with notable amendments passed in 1974 as part of the Watergate scandal aftermath. The Senate Report for the Privacy Act makes the distinction clear:

> [The] term ("individual") is used instead of the term "person" throughout the bill in order to distinguish between the rights which are given to the citizen as an individual under this Act and the rights of proprietorships, businesses and corporations which are not intended to be covered by this Act . . . This definition was also included to exempt (from) the coverage of the bill intelligence files and data banks devoted solely to foreign nationals or maintained by the State Department, the Central Intelligence Agency and other agencies for the purpose of dealing with nonresident aliens and people in other countries.[162]

The Privacy Act's "individual" was narrower than FOIA's "any person" to achieve smaller goals. It lacked the "poetry" of FOIA, which provided rights to anyone anywhere in order to limit power and abuse.[163]

The Privacy Act includes a long list of exemptions. One of the twelve reasons that an agency may share information without consent is "routine use," defined as "the use of such record for a purpose which is *compatible with* the purpose for which it was collected."[164] The agency must state the routine use in the *Federal Register*, where law enforcement databases are also to be disclosed. Although the act directed law enforcement agencies to follow the fair information practices, it also provided another powerful exemption. The act gave agencies the power to make their own exemptions for any system of record used principally for "any activity pertaining to the enforcement of criminal laws."[165] The law was largely bureaucratic, as opposed to legal, and lacked sustained political leadership and pressure. According to the PPSC, which was dissolved in 1977, the Privacy Act "represents a large step forward, but it has not resulted in the general benefits to the public that either its legislative history or the prevailing opinion as to its accomplishments would lead one to expect . . . [and the act] ignores or only marginally addresses some personal data record-keeping policy issues of major importance now and for the future."[166]

The Privacy Act was the closest the US got to comprehensive data protection. After the PPSC presented its report to President Jimmy Carter, he created a task force to develop an initiative that finally resulted in, according to PPSC chair Linowes, the proposal of 200 bills that incorporated the recommendations.[167] These privacy bills were proposed amendments to laws that govern credit, banking, education, and medical care—each with their own consenting characters. They were not computer laws, nor was the

Privacy Act. As the PPSC noted, "The Act's failure to attend to the impact of technological advances on individual liberties and personal privacy is compounded by the manual, or file-cabinet, view of record keeping that underlies it."[168] Without a general law to address computer power and without a data subject, the US set off on a different path that would limit its ability to govern beyond information access and communication privacy.

In France, the Tricot Commission warned that computers would increase power for big governments, cities, and businesses while individuals would feel more helpless and exposed.[169] In response, French leadership took a national survivalist tone that emphasized administrative order. During the 1970s, France remained enthusiastic about *informatique*, executing the second phase of the Plan Calcul across industry, education, infrastructure, and policy. Leaders were politically sensitive to cultural skepticism about computing that might hinder domestic industries and techniques, especially after SAFARI had been portrayed as Big Brother and the American Honeywell company had purchased Bull. The Nora-Minc Report was important to French technology policy but not to the specifics of its first data protection regulations. The report was handed to President Giscard d'Estaing in December 1977 and La loi Informatique et Libertés passed only months later in May 1978. The *Tricot Report* provided the bones for one of Europe's earliest and most influential national data protection laws, at the center of which was the Commission nationale de l'informatique et des libertés (CNIL). Philippe Lemoine was later appointed a CNIL commissioner; Louis Joinet was the CNIL's first chief of staff, and he immediately recruited Marie Georges. Together, they operationalized and evolved French data protection policy, and later European Union data protection law.

Through close oversight from the CNIL, the French built a sophisticated and innovative structure to limit the control that computers might exert over individuals. The law began by stating that computing will be in the service of every citizen and not infringe on human identity or rights. The next article prohibited decision-making on the basis of automated processing of personal information, and the next provided a right to dispute data and logic used in automatic processing. Also prominent was the right of individuals to know about and challenge automated decisions. Public and private sector entities that processed personal information had to declare their plans to the CNIL, unless processing was mandated by another law. Without a strong civil society, French citizens have traditionally trusted their highly bureaucratized and centralized government to protect their interests, and, as such, the data protection law quickly turned to logistics, outlining details of budgets, management, and appeals. This French bureaucracy was to determine morally justified arrangements of computing personal information.

However, buried in article 27 are details about how "*les personnes auprès desquelles sont recueillies des informations nominatives*" ("persons about whom nominative data are gathered") must be informed of automated processing. These details included whether answering was mandatory or optional, what the consequences of not answering were, which physical persons or legal entities would receive the information, and whether there was a right to access and correction. Additionally, article 31 forbade entering or storing in a computer personal data that revealed racial origins, political or religious opinions, or trade union membership. It required *accord exprès* (express agreement) from *l'intéressé* (the interested party). "Consent" (*consentement*) does not show up in French data protection law until 2004, when legislative changes were required by the EU Data Protection Directive of 1995. The French law then refers to the *personne concernée* (person concerned), translated as data subject.

When François Mitterrand was elected president in 1981, he continued the technology policy efforts of his predecessor, declaring, "I will put informatics at the service of men. I will ensure its control by an independent system of justice so that no data bank can serve for the manipulation of the spirit of people. I will inaugurate finally a real redistribution of the means of data processing, which is the condition for an effective redistribution of power."[170] He wanted everyone on earth to have a personal computer, but first everyone in France.[171] As part of the nationalization movement in France, Bull was nationalized by the French government in 1982 and produced machines that could be set up as servers for the revolutionary Minitel system. The effort to pump up the domestic server hardware market was disappointing. Because Bull failed to meet the demand for Unix machines, Minitel services were often set up on American equipment.[172] However, French manufacturers like Telic-Alcatel and Matra made the free Minitel terminals, uniquely transforming French data subjects into computer users at least a decade before the rest of the world (see chapter 3).[173]

Other countries were also looking at national efforts and assessing their own situations. Belgium was considering a registration system and

Denmark was debating rules for different contexts. Luxembourg, the Netherlands, Switzerland, Portugal, and Spain were all discussing regulation, some even considering constitutional changes. The UK and Canada were late to the game, legislatively speaking. The development of Canadian data protection law was a quiet affair. Without any particular scandal or public protests, privacy was included in part IV of the Canadian Human Rights Act of 1977. The Task Force on Privacy and Computers had produced a report, *Privacy and Computers*, in 1972, but it did not call for urgent legislation, instead recommending a privacy commissioner to be appointed in 1977.[174] Aware of Hesse's and Sweden's data protection laws, the Canadian public servants working on the issue did not find a licensing system appropriate, and instead found inspiration in the US Privacy Act. However, an active and designated agency leader was consistently deemed important to the working group, notably because the American system suffered from weak implementation and was vulnerable without such leadership. The first national law to regulate government data use was passed in 1983. As in France and the United States, the Canadian law provided individuals (not yet data subjects) with access to how information was used and with procedures for challenging such use, but consent to acquire or process information is not mentioned until later in the country's data protection history.[175] While the scale of computing in government institutions certainly prompted attention, Canada did not restrict itself to addressing computing or automated data processing.

After the Younger Committee recommended punting the two English bills proposed at the beginning of the 1970s and the Lindop Report demanded action at the end of the decade, the UK did not pass its Data Protection Act until 1984. Even then, it did so dragging its feet. In March 1981, data protection legislation became necessary in order to ratify the Council of Europe's Convention for Data Protection and to comply with the Organization for Economic Cooperation and Development's (OECD) parallel guidelines for transferring data. The writing was on the wall, and nobody wanted to disadvantage UK companies.

The British Data Protection Act required an organization holding personal data for automated processing to register that it was being held and for what purposes with the Data Protection Registrar. "Data subject" was defined as "an individual who is the subject of personal data" and the "data user" was the "person or organisation that 'holds' the data."[176] The codified eight principles were taken directly from the European Council's convention (described in the next section). Like the US, additional disclosures required consent. Consent was otherwise sparsely scattered about in the act for procedures like disclosure in payroll processing.[177] Difficulties passing UK legislation, and establishing a British data subject, reinforced the challenges of passing domestic policy based on an entirely new body of regional law deriving from a recognized but still unformed fundamental right.[178]

TRANSBORDER DATA FLOWS

European countries could not let protections end at their borders, nor could they afford to isolate themselves within a domestic computing market. Schwartz explained, "Permitting an abuse of European citizens' personal information outside of Europe would make a mockery of the decades of efforts expended in creating high levels of protection."[179] Portugal and Spain explicitly required "equivalency" standards of other jurisdictions in order to export data; the French law set forth a way for the CNIL to substantively assess data transfers to other countries; and some countries could simply block transfers. These issues all related to what came to be known as "transborder data flow," defined as "units of information coded electronically for processing by one or more digital computers which transfer or process the information in more than one nation-state."[180] While national data protection of domestic data subjects was certainly one motivation for passing laws, hedging against "data protectionism" was another. There was fear among some working on European economic efforts that national legislation intended to protect "domestic employment, local technology and expertise, home industries, national culture, language, and sovereignty" would serve as "artificial barriers to free information."[181]

At a European transnational level, the Council of Europe's Committee of Ministers concluded that article 8 of the European Convention on Human Rights (ECHR) protecting the right to privacy did not necessarily adequately address concerns such as access to data, the private sector's use of data, or all types of data.[182] It adopted two resolutions (one for private sector electronic data banks in 1973 and one for the public sector in 1974) and determined that a European convention with legal teeth was necessary.[183] It did so when only two data protection laws existed, in Sweden and Hesse. The two resolutions were drafted a year apart, and both recommended laws in the form of basic 1970s FIPs. The Council of Europe took an important leadership role by drafting the Convention for the Protection of Individuals with Regard to Automatic Processing of Personal Data of 1981 (Convention 108).[184]

Like the two resolutions adopted before it, Convention 108 makes little use of consent, including it only once in a minor provision and leaving it without definition or discussion.[185] Although the title of Convention 108 uses the word "individual," within the body it quickly switches to "data subject." The data subject (*personne concernée* in the French translation and *Betroffener* in the German) was defined within the definition of "personal data," which "means any information related to an identified or identifiable individual."[186] Perhaps a bit clunky for resolution titles, the data subject was becoming the operational figure of European data protection.

In 1980, Frits Hondius, who served on the Council of Europe's Committee of Experts on Data Protection, which reviewed Convention 108, explained to an American audience that "computer law" falls into three categories: applying existing law, adapting existing law, and making up new law. "Data protection," he wrote, "represents an example of the third type of legal response to computer problems—creation of an entirely new body of law."[187] Data protection, Hondius concluded, was a "rare example of a branch of law concerned almost solely with a moral issue."[188] The US State Department flatly rejected Convention 108, unable or unwilling to see the moral issue.

The Organization for Economic Cooperation and Development (OECD) saw the international economic potential of computing in the late 1960s and began working closely with a Council of Europe committee to develop what would become the Guidelines on the Protection of Privacy and Transborder Flows of Personal Data. The 1980 OECD Guidelines refer to consent of the data subject specifically in the Collection Limitation Principle, which reads: "There should be limits on the collection of personal data and any such data should be obtained by lawful and fair means and, where appropriate, with the knowledge or consent of the data subject."[189] It is unclear whether and when knowledge *or* consent of the data subject may be appropriate, but paragraph 10 further explains that consent was required (perhaps only) for the disclosure of personal data to a third party. It appears that at this point in time, even the OECD, an economically minded organization, was using consent in terms of fundamental rights as opposed to consumer choice. In reflecting on the motives of the expert committee on the OECD Guidelines, which he chaired and was seconded by Louis Joinet, Australian legal scholar Michael Kirby explained:

> One of the reasons the OECD embarked on its Privacy Guidelines in 1978 was that a gap had opened up in the proper protection of personal data (both automated and conventional). Another reason for these developments was anxiety that differing national legal regulations, superimposed on interconnecting communications technology, would produce serious inefficiencies and economic costs, as well as harm to the value of personal privacy. In this way, the important human right that was at stake was shown to have significant economic implications, deserving the attention of the OECD.[190]

Whereas Convention 108 is legally binding on signators and focused on computing (automated processing), the OECD Guidelines are broad international guidelines supporting a global economy focused on information (personal data). The Council of Europe is not a lawmaking body but can enforce its international agreements through the European Court of Human Rights. The OECD referred to data subjects beyond their commercial identity but was not in the business of international rights and was not necessarily talking about the act of computing (though the focus of its report was on information technologies). Appropriately, the OECD Guidelines seem to be a muddled negotiation of the state of consent to compute at the time.

A small number of experts emerged to provide insights and guidance. European practitioners like Simitis and Hondius published extensively on "data protection law" in the 1980s and 1990s. Others began to refer to themselves as experts in European "computer law."[191] European law schools offered "computer law" and "informatics law" courses beginning in the 1980s. In Canada, David Flaherty wrote extensively about comparative surveillance and served as data protection registrar in British Columbia. In 1981, Jon Bing, a prominent Norwegian legal scholar who had led the Council of Europe's Committee on Legal Data Processing, drafted the OECD Bing Report, which distinguished computerized data from its transmission. In the US, "computer law" did not emerge, but later cyberlaw would in response to the internet (discussed fully in chapter 3). US scholars writing on the same subjects as computer law and data protection law were often referred to as international legal scholars, political scientists, or sociologists with an interest in privacy. This small cohort was actively engaged in policy work, providing reports, expertise, and internationally minded, comparative insights over decades. Political scientist Priscilla Regan wrote about British and American privacy and was one of the earliest voices to call for a recognition of the social (as opposed to individual) value of privacy.[192] Sociologist James B. Rule's *Private Lives and Public Surveillance* observed two worlds in 1974: the "experiential reality" and the "paper world," warning about the latter's growth and our potential reliance on it.[193] Although none of these scholars were computer scientists, Joel Reidenberg, a prescient, prolific, and preeminent scholar in both cyberlaw and privacy law, was a graduate of the Dartmouth program, which required all students to learn BASIC in their entry math courses, a skill that directed his early legal career while working in France.[194] Reidenberg worked with international legal scholar Paul Schwartz to translate American approaches to the new European Union and vice versa in the 1990s.[195]

Within the European Union, the European Parliament had been pushing for rules protecting against "computerized retrieval of personal data" since 1979 but could not rally the European Commission.[196] The Commission prioritized European developments of information and communication technologies and recommended that Member States ratify Convention 108 knowing that it was only a series of guidelines not intended to substitute national laws. Convention 108 also included an article that stated countries should not obstruct transborder data flows for the sole purpose of protecting privacy. The Commission was not focused on rights, but rather on a European "information market."

The European integration that had begun in the 1950s and the 1957 Treaty of Rome (more formally, the Treaty Establishing the European Economic Community, EEC) waned with economic decline in the 1970s, but was revitalized in the 1980s by leadership focused on market reforms. The Single European Act of 1987 was the first major revision of the EEC; it was dedicated to establishing an internal EU market by the end of 1992. Because little legal harmonization occurred under Convention 108 and the OECD Guidelines, the Commission was justified under these new political circumstances in proposing a directive "to ensure the establishment and functioning of the Internal Market," under the power granted in article 100a of the EEC. The Commission issued the draft of the directive in September 1990. Parliament heavily amended it in March 1992 and added "Free Movement of Such Data" to the title months later. Then it was heavily negotiated with the Council for two years.

In 1990, concerned about market disruption deriving from the conflict between national data protection regimes within the European Economic Community (at the time, there were only twelve members and seven had data protection laws), the European Parliament prompted the European Commission to address "protection of individuals in relation to the processing of personal data."[197] The 1992 Treaty on European Union (Maastricht Treaty) initiated "the most profound structural change in the history of the European Community—the transition from a merely economic union to a decidedly political one."[198] The Maastricht Treaty set the European Union on an alternative path dedicated to the common abolition of internal barriers, citizenship, money, foreign security, and justice. The explicit commitment to the furtherance of democratic societies was the product of Member States' emphasis on safeguarding fundamental rights beyond the decisions of the European Court. The Commission changed positions and took up a rhetoric wherein processing personal data would include its impact on fundamental rights.[199] The Commission need to provide not only protection but a "high degree" of protection, which is the strongest form expressed in the EU.[200]

Passed in 1995, the Data Protection (DP) Directive was a long-negotiated document. Simitis recalled, "The primary interest of the Member States was not to achieve new, union-wide principles, but rather to preserve their own, familiar rules. A harmonization of the regulatory regimes [was], therefore, perfectly tolerable to a Member State as long as it amounts to a reproduction of the State's specific national approach."[201] The British and Dutch encouragement of codes of conduct was incorporated. The French duty to notify a data protection agency when processing of personal information would be wholly or partly automated was included. The special prohibition on processing sensitive data derived from French, Spanish, and Portuguese laws was considered, but it conflicted with the German approach articulated by Simitis that treated data as contextual, not categorical. Joinet presented a compromise: all personal data would be covered, but sensitive data would be given special treatment. Most prominently, the Directive required Member States to pass laws that included a predetermined justification for computing personal information, the essence of all German data protection laws and those with licensing regimes. In other words, the act of computing required a moral and legal transformation between parties: a legal basis. Those parties had names like "data controller" and "data processor" and "data subject," along with definitions. European Union documents are translated into all official member languages. In the Directive's definition of personal data, the German *betroffene Person* and French *personne concernée* (both earlier translated to "person concerned") are translated as "data subject."[202]

Unlike in the US, a data subject's rights are infringed if personal data is collected or processed without a legal basis. The data subject's consent provided an important legal basis. The five other legal bases were: 1) to fulfill a contract with the individual; 2) to comply with a legal obligation that requires the processing of personal data; 3) to serve vital interests (e.g., when processing data is necessary to protect someone's life); 4) to complete a public task (e.g., those tasks assigned to government departments, schools, hospitals, etc.); and 5) to achieve legitimate interests of the processor (other than public authorities), not outweighed by negative effects to the data subject's rights and freedoms. The first three are presumed to have already met a justification for consent either by assumption (fulfillment of contract or physical health) or societal recognition (public tasks), but other "legitimate interests" were less clear. After the European Parliament amended the European Commission proposal, consent was moved from article 12 under data subjects' rights to article 2 under definitions but was not given priority as a ground for processing data. Under the 1995 DP Directive, consent was required both as a precondition for processing in article 7(a) and for processing sensitive information (article 8(2)(a)). Georges remembers spending very little time on the topic of consent, which needed to be written at the time.[203] Presumably, most fair processing would fall easily into one of the other legal bases.[204]

Although it passed in 1995, arguably the year the web came on most people's radar when a number of major players came online (including Match.com, Amazon, eBay, Yahoo!, and Craigslist), the DP Directive wasn't about the internet.[205] As the Directive was being debated and drafted, the internet was still only used by a few people. Estimates in 1992, for instance, were around four million users worldwide.[206] The Symposium issue of the *Iowa Law Review* from 1995 contains articles by some of data protection's most prominent figures, like Simitis, Reidenberg, and Schwartz. None of them mentions the word "internet." Schwartz explains in his first footnote what kind of transborder data flows were common and relevant to the Directive: (1) personnel departments; (2) banks, insurance companies, credit card companies, and credit bureaus; (3) direct marketing; (4) airlines, travel agencies, and other businesses involved in tourism; (5) companies that seek to deliver goods to or otherwise trade with international customers; and (6) within the public sector: police, customs, tax departments, and public pension agencies.[207] No internet.

By the mid-1990s, many individuals operated computers either at work or at home, but the transborder data flows common at the time still involved data created, input, and shared by another person, not the data subject. A decade earlier, the *New York Times* reported that the US had passed a milestone: desktop computer sales surpassed mainframe machine sales. But even in the mid-1980s, it was still obvious that "thousands of big corporations" would "still need a complex, powerful mainframe computer with enough electronic memory and circuitry to store and rapidly process the huge amounts of information that must be marshaled to print out the nation's monthly Social Security checks, or a major corporation's payroll, or to keep track of a multinational's worldwide inventories."[208] But for smaller computer tasks, a smaller machine would suffice. The Seven Dwarfs (by this time referred to as the BUNCH after mergers dropped the number from seven to five) tried to get in on the microcomputer by looking to Japan for new products and parts but were too late (see figure 2.2). A fresh set of players had arrived: Apple, Tandy, and Commodore. But IBM had legitimized the personal computer when it entered the market in 1981, still investing heavily in its profitable mainframe lines throughout the 1980s and early 1990s.[209]

The European microcomputer industry lagged but was viewed optimistically as steady and reliable, unlike in the US. "In the US, everybody went crazy with computers . . . In Europe, it's been a normal business from the beginning," explained Howard Elias, Tandy Europe's marketing director.[210] He further explained that European markets were not considered as easily computerized: "In the US 50 years ago [in the 1930s] you had the corner butcher, the corner baker, and the corner candlestickmaker. Now they've developed into the supermarket, and the department store. But in Europe, even in the major cities, they've still got the corner butcher and so on. And part of that has to do with the way society is organized."[211] It was not just cultural differences and tradition; software at the time was not created for most European languages. Microcomputer penetration into white-collar work forces was predicted to reach 20 percent by 1990, up from about 3

FIGURE 2.2
Editorial illustration by Randy Jones, in David E. Sanger, "Bailing Out the Mainframe Industry," *New York Times*, February 5, 1984, sec. 3, 1. Reproduced with permission from the artist.

percent in 1985.[212] IBM also dominated the personal computer market in Europe, but Apple and other American companies like Hewlett Packard and Compaq competed with, at least for a time, ICL in Britain, Siemens Nixdorf in Germany, and Bull in France for market share.

More and more people were being exposed to computers, but few people entered their data into systems, more often using programs at work for administrative tasks or at home for fun. Although the emerging computer networks described in the next chapter served as the transportation channels for much of this data, it may also have been shipped as magnetic tapes or punched cards across borders. The data subject may have had increased choices given the growing use and contexts for computer data, but would have remained distant from the data collection and processing occurring while the DP Directive was drafted. A network of managers, clerks, and technicians with complicated and varied roles and politics stood between the data subjects and their data. As such, the DP Directive was drafted not for the internet of today, with personal users directly inputting their own data, but rather to support a discourse of control around transborder data flows based on investments in national computing that needed to be internationally networked (see chapter 3).

The US never embraced the data subject, instead focusing on an existing cast of non-computer characters like the student and the patient. US law offered legal protection from computing power within the contexts of its institutional use. It passed the Fair Credit Reporting Act for credit bureaus and consumers, the Family Educational Rights and Privacy Act for schools and students, and the Video Privacy Protection Act for "video tape service providers" and movie renters, to name only a few laws that govern personal data in the US. Nonetheless, its missing data subject caused problems as transborder data flows and early computer networks turned into dominant global telecommunication systems. In the years following the US Privacy Act, no comprehensive data protection law was passed. By contrast, the data subject was a vital part of Europe's legal approach to computing as a morally transformative act. Across Europe, the data subject came to be recognized, and an entire intricate global body of data protection law was created from the above national efforts.[213] When the European Union created a set of fundamental rights at the end of the twentieth century, it included a special right to data protection for data subjects. Without a similar construction of computer law or data protection law, and without a data subject in the US, harmonization or interoperability have proven to be challenging and resulted in cookie notifications as consent theater.

3 NETWORKING THE USER

The origin story of the user, the second computer character central to cookie consent, comes from communication privacy law. It is a story that begins with early efforts in different countries to build computer networks to facilitate communication. Minitel, Prestel, BBSes, AOL, Usenet, ARPANET, and Euronet are just a few networks that preceded the now dominant Silicon Valley platforms of Web 2.0.[1] These networks are layers and layered and connected and disconnected over the decades. Their particular structures, economic arrangements, and identity strategies have contributed to the legal construction of users within national legislation passed to update communication privacy and criminal laws. The user in these national laws would later be folded into technology policy debates and shape the transatlantic distinctions around cookies.

Over the course of the 1970s and 1980s, the Cold War mainframe paradigm in which central automated control reigned gave way to a growing community of microcomputer hobbyists driven by exploration and discovery.[2] Throughout this transition, the engineer/programmer/developer made space for the user and the hacker.[3] Tech writer Joanne McNeil points out one of these distinctions in the introduction to her book *Lurking: How a Person Became a User*. She notes that while plenty of people find "user" distasteful, it is a useful label because "people" hides that the internet is made up of two classes of people: "users" and "developers." Developers, in McNeil's assessment, are not driven by the hobbyist's need to tinker or showcase cleverness, but by human–computer interaction principles, corporate engagement mandates, and the attention of venture capital.

Anthropologist Gabriella Coleman tracks the hobbyist-turned-hacker in her ethnographic work, uncovering the liberalism, pleasure, and poetics of hacking defined by values of freedom, privacy, and access.[4] As such, hackers too became another class distinct from the user and even the developer. Still, the lines are fuzzy. The Jargon File, a slang usage dictionary for programmers published in 1983 as *The Hacker's Dictionary*, defined the user in relation to the hacker: "User: Someone doing 'real work' with the computer, using it as a means rather than an end. Someone who pays to use a computer." The entry goes on to explain, "The general theory behind this term is that there are two classes of people who work with a program: there are implementors (hackers) and lusers. Users are looked down on by hackers to some extent because they don't understand the full ramifications of the system in all its glory."[5] Note: a luser is "a user; esp. one who is also a loser."[6] The politics of legitimate developers versus illicit hackers is tied up in the politics of anonymous users versus lame lusers, and laws engaged and shaped these politics in important (and sometimes odd) ways.

Through iterations of communication innovations between the 1960s and 1990s, the user became legally defined in two areas of law: communication privacy law and computer crime law. Across countries, communication laws were consistently updated to protect users from both "electronic eavesdroppers" and hackers. In countries that were trying to grapple with old customs of interception, promote a particular vision of domestic computing, and support international agreements, these updates to communication law could be quite awkward. Users were also distinguished from hackers in computer crime laws that passed in the same period. Both forms of legislation prohibited access to computer communication without authorization, which often appeared in the language of or interchangeably with consent. Whereas consent from the data subject in data protection law is all about control, consent from the user in communication privacy law is all about authorization.

AUTHORIZED USERS IN THE PROTOCOL WARS: DATA NETWORKS IN THE UK

Users on computer networks are evolutions of users on prior international communication networks. The United Nations and other political and thought leaders pushed the free flow of information after World War II to promote peacekeeping efforts, strong democratic societies, and economic growth. As data became understood as valuable to those beyond the credit reporting and intelligence worlds in the 1970s, it was shipped across borders in paper formats like files and punched cards as well as tape and disks, but states were investing in telematics and beginning to lay down or update networks to transfer the information electronically.[7] The discourse around telematics became one of individual liberation and global harmony, with a few exceptions. This discourse was complicated, however, by a growing skepticism of European integration over the period.[8] The strong support of the Marshall Plan, which provided US aid to Western Europe after World War II, waned during the Cold War. Although European postal, telegraph, and telephone corporations (PTTs) had always resided neatly within the state, cross-border connections were established through agreements and coordination with public and private purveyors of communication services.[9] Computer networks were developed by and around government agencies to serve numerous domestic needs, the articulations of which revealed skepticism of US-European relations and prioritized competition for global relevance.

New routing methods for telecommunications messages generated major interest and controversies in computer networks across universities, industries, and governments. Paul Baran (who, as discussed in chapter 2, worked at the Rand Corporation in the United States and testified before Congress about the National Data Center) and Donald Davies (who worked at the British government's National Physical Laboratory, NPL) both independently pursued packet switching as a method of transporting and routing information across networks. Packet switching is a dynamic form of data transmission: a message is broken into pieces and grouped into packets, which are then sent over a digital network.[10] Throughout the late 1960s and early 1970s, Davies saw more success than Baran at promoting packet-switched networks and inspired their use in the US ARPANET, the pilot network project under which the Advanced Research Projects Agency (ARPA) planned to link various computer science projects it sponsored in 1967.[11] University College London's connection was the first British node of the many single transcontinental ARPANET nodes, but Davies's pioneering NPL network followed.[12] ARPANET is often considered the network that turned into "the" internet, but it will get company in this book's history of the user.

In her study of the Requests for Comments (RFC) filed over ten years (1969–1979) of ARPANET internet construction, communications scholar Sandra Braman found privacy addressed at the network, host, and user levels.[13] At the network level, privacy and authentication of users accessing sensitive research or data on a network was decentralized in order to be robust and reliable. Although most of the RFCs were signed using full names and affiliations, there was an expectation that users on ARPANET would or could be anonymous and that log-ins with passwords would function as means of access authentication. In addition to sensitive information, remote users needed to be authenticated for billing, because the economic burdens on hosts could potentially be crippling. Communications historian Rita Zajácz suggests the network achieved anonymity somewhat by accident, as opposed to by explicit political design. She notes that trusted people oversaw the secure locations that ran each of the sixty computers on the network. If anything went wrong or suspicious activity occurred, it would have been easy to figure out the source by investigating the header information.[14]

This anonymity-after-authorization was present on some European networks as well, but under different circumstances. The European Community (and later Union) approach to developing internet protocols was "somewhat reluctant," because several countries were pursuing their own network protocols, which were unique to who was developing them and why.[15] An important distinction between the US and Europe is the role of the PTT. Whereas the internet in the US was funded by and in coordination with the government, in Europe it was funded by PTT corporations that were embedded in the state. PTTs operated as centralized, state-administered monopolies, and international connections were established through agreements and managed by an international body, later the International Telecommunication Union.[16] While the PTTs in some countries supported research efforts to develop packet-switched networks, they were also developing their own networks.[17] The PTTs were protective of the standards used on their networks and would shape European policy as the region navigated the convergence of computing and communications. The PTTs also represented the kind and power of entities that would potentially have access to information about people using the networks.

For European networks, PTTs wanted to expand their existing business and favored "virtual circuits." In a virtual circuit, a route is established, data is broken into packets, and the packets are sent down the route in order. STS scholar Andrew Russell explains that the design of virtual circuits "mimicked circuit switching's technical and organizational routines" and also expressed a form of protocol protectionism familiar to the European telecom industry.[18] Virtual circuits are analogous to telephone services: a connection is established and messages are delivered in the order they are sent. Alternative protocols developed in research institutions broke data into packets, marked each with the destination information, and sent them along the network using whatever route was the most efficient, to be reassembled upon delivery. These were often referred to as *datagrams*. Datagrams are analogous to postcards: each packet is marked up and sent separately, and the receiver must collect and organize all of them into the original message. In a competition that came to be known as the Protocol Wars, technopolitical actors debated whether virtual circuit or datagram standards provided better reliability and efficiency.[19] The virtual circuit was the basis for the X.25 standard championed by European PTTs. X.25 competed with others, including datagram-inspired standards like TCP/IP (the one ARPANET implemented in 1983 and that we still use today).

The French battle of the Protocol Wars was representative of the players and politics. The French telecommunications authority's research arm CNET began developing Réseau à Commutation de Paquets (RCP), which worked with virtual circuits and added durability to an experimental network that the PTT was creating called Transpac.[20] The French PTT project was being built using the X.25 standard on the American Digital Equipment Corporation (DEC) computers.[21] This was distinct from what was being developed at the French research network Cyclades. At Cyclades, pioneering computer scientist Louis Pouzin made significant advancements in the development of the datagram packet-switching concept, which was promoted as "connectionless" and was built using French CII computers. In an interview with Russell, Pouzin further explained, "That means you have no relationship established between sender and receiver. Things just go separately, one by one, like photons."[22] The Cyclades team was the first to implement Davies's end-to-end principles, as they would later be called. The hosts, not a centralized network service, would be responsible for successful transport of data. In other words, there was less authority involved in datagrams versus virtual circuits. The French PTT declined to support the datagram. The Cyclades team was told the future was virtual circuits and the X.25 standard. Its funding was pulled, and the researchers moved onto other projects. Pouzin and his team spoke out against what they saw as a financially and politically motivated rejection of datagrams to "keep a firm grip . . . on users."[23] Although the ARPANET and Cyclades were never connected for political reasons, the postcard-like protocol and end-to-end principles would be incorporated into the ARPANET, and Pouzin would later say the Americans saved the datagram.[24]

In the UK, the Post Office was the first to announce the development of a national packet-switched network for the public in 1973. The Post Office was initially skeptical but was nudged by those in London's Real Time Club, a dining club established in the 1960s by entrepreneurs and computer enthusiasts that took up lobbying causes promoting British technology developments.[25] The Department of Industry confirmed interest from British companies and, using British Ferranti computers, launched the Experimental Packet Switched Service (EPSS) in 1977, which was turned into a public network called Packet Switch Stream (PSS) in 1980.[26] As Davies and the NPL continued to connect to computer networks, they prioritized European networks like the EPSS and the European Informatics Network (EIN), because the UK was trying to join the European Community at the time.[27] The NPL and Department of Industry envisioned a business-oriented network that would limit user exposure to computer complexity, save computing costs, and increase network efficiency. Privacy on such a network—on any network, really—had only recently been interrogated in the UK.

Despite its use in the 1800s, wiretapping received close regulatory scrutiny only in the late 1950s. In 1957, Lord Norman Birkett was tasked with chairing an interrogation into wiretapping following the Marrinan Affair of the previous year, in which the police were granted authority by the Home Secretary to wiretap a prominent gangster who was talking to a lawyer, Patrick Marrinan, who was under investigation by the bar.[28] The Home Secretary gave the bar investigators copies of the recordings of their conversations. The public and the political opposition demanded an inquiry. Reactions from the press ranged from calls to ban the practice entirely to comparisons like, "Very little difference [exists] between having a man trailed by detectives and having his telephone tapped."[29] The Birkett Committee reflected that the apparent legality of wiretapping was perhaps because the power to intercept communications originally laid with the Crown, which did so according to its prerogative. The committee concluded that wiretapping phones was well within the Crown's long-recognized power to intercept communications generally. This power emerged with the British Post Office, which generated revenue but also served as the government's eyes and ears, with post office officials acting as intelligence agents.[30] In 1660, parliament established a post office, and the first postmaster general allowed inspection by the secretaries of state at their discretion.[31] Beginning in 1710, written warrants from secretaries of state were required, and anyone who abused their power to open or hold letters was fined.[32] These warrants would often include a simple request to hold and read all letters to and from a long list of individuals.[33] Later, one of the secretaries contacted the head of the Post Office directly to tap phones. The Postmaster General and the Home Secretary assessed the arrangement in the 1930s and decided that the Home Secretary should add a layer of authority. In light of the Marrinan Affair, the Birkett Committee recommended that no interception by the Home Secretary ever be disclosed "to private individuals or private bodies or domestic tribunals of any kind."[34] But the committee was otherwise satisfied that British wiretapping was exclusively in the hands of Secretaries of State for national security. It found that adding independent review from magistrates or high court judges would add no improvement and was quelled by the Home Office's strict practice of never introducing wiretap evidence in criminal prosecutions—and never confirming or denying that tapping had occurred.

The practice of tapping was at the heart of *Malone v. UK* in 1984. The Post Office Act of 1969 had moved the Post Office out of the State Department, but the Home Secretary still issued wiretap warrants and had issued one to tap James Malone's phone. The recorded conversation was inadvertently provided to the court as evidence that Malone had stolen property. While Malone waited for a retrial, he filed suit against the police commissioner in 1979. He sought to have the interception declared unlawful, because he had not consented and so suffered a violation of his privacy rights, as articulated in article 8 of the European Convention on Human Rights. The judge dismissed Malone's privacy claims because no trespass had occurred and found that article 8 conferred no rights on UK citizens (Parliament should draft a statute to change that, he said). Malone submitted his issue to the European Commission, which then referred the case to the European Court of Human Rights (ECtHR) in 1983.[35] The ECtHR found Britain's interception arrangement violated article 8 because it was not in accordance with a satisfactory law. It elaborated on the requirements of article 8: interception of communication must have a basis in domestic law, and that law must be accessible and written clearly.

In 1985, the UK Parliament passed the Interception of Communications Act, which made it a crime to "intercept a communication in the course of its transmission by post or by means of a public telecommunication system."[36] A person was not guilty if the interception was performed in accordance with a warrant issued by the Secretary of State, or the sender *or* receiver consented to the interception.[37] In other words, when the UK finally passed laws articulating the conditions for tapping communication, it required either a state authorization or authorization from any party to the communication. European policy similarly motivated UK computer crime legislation. Policymakers did not enact the UK legislation until 1990, after the Council of Europe's Recommendation No. R (89) 9 called upon members to harmonize their computer crime laws and after the British public became increasingly concerned about reports of communication disruptions and damage to computer files. The law looked much like the other national laws passed before it, relying on "authorization" to determine legal computer use.[38] The law criminalized actions such as unauthorized access, modification, and destruction of data. Pad and Gandalf, two hackers who made up the group 8lgm (Eight-Legged Groove Machine), were the first people convicted under the law in 1993. The pair, whose real names are Neil Woods and Karl Strickland, had not met in person before they were arrested, only on computers connected to their phone lines.[39] They were jailed for "intellectual joyriding," as the judge put it, in the form of unauthorized access to computers all over the world, and sentenced to six months in prison.[40]

ANONYMOUS USERS ON MINITEL: VIDEOTEXT IN FRANCE

Videotex, another iteration of user technology, developed in the UK and was later taken up in France. It failed to gain traction in the former but became revolutionary in the latter.[41] Videotex was a data retrieval system structured like a library or catalog, with a vast number of files held in memory on a host computer that are transmitted over telephone or cable television lines accessed by a terminal (often a television set with an adaptor) using a modem. Menus organized the actual "information frames," which were provided by wire services, newspapers, magazines, retail stores, advertisers, government agencies, financial companies, and entertainment firms. Information frames relating to sales were provided free of charge, but others required a fee per page. Starting in 1971, the British Post Office began developing its videotex system and launched it in 1979 as Prestel. It ran on the public PSS network. To gain access to the Prestel information retrieval computer, a user would have to establish a connection, dial the Prestel number, type in an identification code, and then provide a password. When a user wanted to buy a product or service, a special frame was selected with fields indicating the product or service, the name and address of the user, and the user's credit card information. These message frames were stored on central mainframe computers and later retrieved by the information provider. For billing, Prestel kept records of royalties owed to an information provider, number of frames selected, and connection times. By 1981, it was offering a database of more than 150,000 frames provided by 150 information providers who created the frames and made them available to the "UpDate Computer."[42] The British Post Office was toying with the idea of providing worldwide service, but any such plans were disrupted by dramatic privatization and liberalization of the telecom industry in the UK. For the most part, videotex networks failed, except in France.

In the late 1970s and early 1980s, France designed and implemented a plan to revitalize its deteriorating telecommunications infrastructure: Minitel. Media scholars Julien Mailland and Kevin Driscoll present many reasons for Minitel's meteoric rise in their book *Minitel: Welcome to the Internet*, but the most important may have been that each French citizen who owned a phone line was given a Minitel terminal for free.[43] Via their Minitel terminals, users connected to the centralized Transpac network, which was publicly built and maintained. Each terminal was made up of a display, keyboard, and modem that a caller could use to dial up to a *point d'accès vidéotex* (PAVI), which served as a gateway to a directory of known Minitel Services. The user would then type in which service they wished to visit, at which point the switch created a virtual circuit over the public data network, and data could flow back and forth between the terminal and house server.

Minitel was unique from other international X.25 or videotex networks in that it allowed private service providers on the network to own their own servers. France Telecom oversaw only the network upon which

Minitel operated using this unique version of the X.25 protocol. However, the Minitel implementation wouldn't allow servers to connect directly to one another, effectively creating a centralized communication structure for hosts.[44] Despite the centralization of the data flows, users still enjoyed anonymity, as all Minitel connections were anonymized. No usernames or passwords were required. A hallmark of this anonymity appeared on Minitel's centralized service marketplace, or *le Kiosque* (the Kiosk), where users could access private services like banking, travel arrangements, message boards, and games. Having a centralized marketplace in which users were charged for connection time allowed the traffic to be anonymized for both the service provider and the caller. Users punched in a four-digit code and the name of the service they wanted, but France Telecom billed only for time spent on the system. Specific service providers did not need to keep track of visitors to bill them because they simply got a cut of fees (about 60 percent) paid to France Telecom.[45] The sex chat services, referred to as "pink Minitel," were the most prominent on the Kiosk and represented the anonymous culture of Minitel. Both caller and service provider enjoyed anonymity as "users don pseudonyms like Noir, Phantom and Sex Fiend, in order to flirt, simultaneously and anonymously."[46] On the service provider's side, the demand for pink Minitel created "animatrices, a new type of information worker whose job was, in the words of one popular song of the period, to 'digitally undress' users."[47] (See figure 3.1.)

These anonymity structures were bolstered by legislation like the Postal and Electronic Communications Code, which criminalized the violation of secrecy in correspondence by those entrusted to provide telecommunication services.[48] Correspondence privacy law in France is old and complex. King Louis XI transitioned France away from private postal carriers to become the first European country to nationalize the postal service in 1464. The post and telecommunications technologies that followed were the exclusive task of the state. French constitutions, of which there have been fifteen since the first in 1789, included implicit rights to communication privacy in the right to freedom of expression.[49]

Nonetheless, wiretapping in France existed in a precarious legal state until 1970, because, while an earlier court case seemed to categorize the act as prohibited, it had not been declared so. In 1953, a judicially sanctioned investigation into bribery charges convinced one of the victims to telephone the suspect and ask incriminating questions while the investigators

FIGURE 3.1
Editorial illustration, Justine De Lacy, "France: The Sexy Computer," *The Atlantic* 260, no. 1 (July 1987): 18. Reproduced with permission from *The Atlantic*.

listened on an extension phone. The Court of Cassation, the highest court in the French judiciary, overturned the resulting conviction on grounds of general fairness and rights due the defense.[50] The consensus was that surely wiretapping would also cross the line as a sneaky trick, but the court did not take up the question directly. Instead, the legislature enacted the law On Reinforcing the Guarantees of Individual Rights of Citizens in 1970.[51] In it, article 368 declared that anyone who intentionally infringed upon the intimacy of private life by "listening in on, recording of, or transmitting of, by means of any device, the words said in a private place by a person, without the latter's consent," would be punished. Article 368 did not make exceptions for anyone, not even law enforcement, but remained limited to intentional invasions into intimacies of private life.

However, in 1980, the Court of Cassation found no error was made by a magistrate who allowed evidence gained by tapping a suspect's phone, a decision it considered to be in line with the court's acceptance of seizure of letters from the postal service despite there being no explicit exception in the criminal code. When police used wiretapping as a preliminary investigative tool to confirm suspicions, however, the Court of Cassation in 1989 decided the invasion was a violation of article 8, as well as French criminal code, despite the other party's consent to the police recording. It was not until 1991 that France enacted a wiretapping statute to explicitly manage the intrusion from public authorities. The Wiretapping Act controlled and legitimized wiretaps performed by judicial investigations by requiring written authorization that demanded identification of the line to be tapped, the offense justifying the tap, and the duration. Magistrates could, however, order wiretaps wherever and whenever they perceived one necessary to an investigation.[52]

Processes for authorized interception by the government were put in place three years after France criminalized unauthorized computer access by others. In 1988, France passed specific computer crime laws that addressed techno-enthusiastic hackers as well as anti-computer groups. While the difference between hacker and user/luser is transnational, the characteristics of that difference are not universal. European hackers fought to shape the computer systems around them, but this often also meant hacking the "Americanness" of the equipment to meet instrumental ends.[53] From changing language settings and keyboard standards to altering plugs and soldering parts, European users tinkered with their American products to get computers to meet their intended goals in an unimagined context. In the early 1980s, a group called the Committee for Liquidation or Subversion of Computers (Comité Liquidant Ou Détournant les Ordinateurs, CLODO) attacked computer company facilities, setting fire to a Sperry Corporation office in Toulouse.[54] After CLODO set fire to CII-Honeywell-Bull, they sent a letter to *Libération* that explained, "We are workers in the field of data processing and therefore well placed to know the current and future dangers of data processing and telematics. The computer is the favorite tool of the dominant. It is used to exploit, to put on file, to control, and to repress."[55] John Badham's techno-thriller *WarGames*, in which Matthew Broderick plays a hacker who accesses a military supercomputer with the capacity to start nuclear war, was released in France in 1983, prompting the satirical newspaper *Le Canard enchaîné* to publish details, in all seriousness, about how its journalists had used Minitel to access computers at Mururoa nuclear base.[56] Soon after, French computer crime laws were among the earliest to pass in Europe. The 1988 law criminalized fraudulent intrusion in, access to, and destruction of automated data processing systems, activities performed by hackers not users.[57]

CCC HACKERS AGAINST BTX: BBSES IN GERMANY

Bulletin board systems also developed parallel to these other network initiatives. BBSes served as a community hub for a user who was willing to tinker, and the technology attracted particular politics. Although BBS, like all the networks discussed, were transnational, the German BBS was home to the Chaos Computer Club (CCC), a politically driven group of computer hobbyists/hackers that was started in 1981. The group was also inspired by *WarGames* and accounts of American hackers, but it gained so much fame and legitimacy that it is now a major political force in European tech policy. BBS was dreamed up, designed, and built over thirty days in Chicago at the beginning of 1978 as a form of networked communication that allowed anyone with a terminal or computer equipped with a modem to call in and leave or retrieve messages.[58] The actual information presented on the bulletin board was hosted on the system operator's (sysop) computer. As long as the sysop's computer had a display, keyboard, modem, and connection to a phone line, other computer hobbyists could dial into their BBSes. BBSes ranged from basic message boards to sophisticated offerings of software, multitopic forums, emails, and games.[59] Unlike commercial online services, almost all bulletin boards were nonprofit, and many charged nothing.[60] The major cost to the BBS caller was that of using a phone line to dial into a BBS, so initial BBSes created local networks to avoid the cost of long-distance calling.

In a 1989 how-to article in the *Washington Post*, the author stresses that, beginning with a user's first call into a BBS, real name policies are the norm. Providing real names and ensuring the identities of BBS callers was a way to establish trust with the BBS community. The article warns, "Some system operators will ask you to register, giving your name and address, and a telephone number they can call to check you out. The purpose is to weed out users who only want to cause mischief. This sort of registration procedure is a sign of a responsibly run board and you should go ahead and sign up."[61] According to communications researcher Kevin Ackermann, to be anonymous on a BBS often suggested that you operated in an ethically gray area, at least on BBSes not dedicated to network tinkering.[62] Anonymous individuals on BBSes were often associated with pirates or "phreakers"—people who used other peoples' credit card numbers to pay for long-distance connection costs.[63] Anonymous users could operate on BBSes run by a sysop who simply asked for no self-disclosed identification. However, the lack of a self-disclosed identity did not mean that the anonymous user was impossible to track down if legally necessary, of course. When two teenage boys used someone else's log-in to a national BBS and sent sexually vulgar messages to female classmates, they were unmasked through a "high-tech tracing job by the telephone company."[64]

Bulletin board software gave the people who ran them broad power over all parts of the operation. The sysop controlled all information and authority in a BBS and largely dictated identity and privacy practices. Additionally, anything that a user uploaded or posted to a local bulletin board system was stored on the sysop's computer. If a wily user obtained the privileges held by a system operator, which was as simple as figuring out a sysop's password, they could tamper with any of information in the system.[65] A hallmark of the BBS era of networked computing was the uniqueness of each network. In his assessment of content moderation practices on BBSes, Driscoll notes that BBSes were sites of experimentation for community management.[66] Because BBS sysops had total control over the rules and standards of operation on their bulletin boards, sysops bucked the developed norms of BBS usage whenever they saw fit. For example, identity verification was an established norm, but if a sysop wanted to reject it, they had complete authority to create a BBS in which the only rule was that real identities could not be shared.

To overcome the prohibitive costs of long-distance BBS calls, artist and techie (among many other things) Tom Jennings built FidoNet, a store-and-forward decentralized network of bulletin board systems.[67] FidoNet managed dialing and connecting to other systems, and did it in the middle of the night, when rates were the lowest. Jennings's network was originally software that would link together BBSes using Jennings's BBS software, but later versions of the network operated with different BBS software. To operate on FidoNet, sysops only needed to have a copy of Jennings's FidoNet software, a list of all the network addresses of other nodes, and an individual node address for their own spot on the network.[68] There was an element of collaboration and benevolence to the operation of the network, because sysops were volunteers who made their nodes available for the benefit of expanding the network. Eventually, the network grew large enough that different organizational committees were set up to ensure new software would be compatible with the network and to interface with modem manufacturers.[69]

FidoNet systems provided international BBS access to certain users in Europe, while nearly every European PTT also operated its own version of a videotex or teletext system. In Germany, this system was Bildschirmtext (BTX), run by the German post office. BTX became a target for the CCC. The Deutsche Bundespost held a monopoly on all mail, telephone, computer networks, and hardware, a regime which the CCC openly rejected and creatively tested. For instance, the Telecommunications Device Act prohibited the use of tools like modems without official postal approval, so the CCC distributed blueprints of non-approved modems (see figure 3.2). In 1977, the Bundespost contracted with Britain to bring Prestel to West Germany, but it wasn't until 1980 that technical trials began with the rollout of about 6,000 terminals. BTX was heralded as an icon of modern German telecommunication. Yet when it was introduced in 1984, the CCC immediately opposed it, declaring that the centralized system granted the post office full control over users and transformed them into passive consumers. BTX distinguished between subscribers and content providers, and each was given unique equipment. Subscriber terminals could only download and render

FIGURE 3.2
Cover of *Die Datenschleuder* (The data slingshot), the irregular publication of the Chaos Computer Club, issue no. 30, September 1989. The dialogue box reads, "may be opened by the post office for testing purposes."

files, but provider terminals could edit. However, upon examination, the CCC discovered that the only difference between a very cheap subscriber device and a very expensive provider device were two buttons that could be added by drilling two holes in the case. The CCC preferred the design and culture of BBSes. BBSes were the "digital counter public," where German computer enthusiasts found members of the CCC and soon became a target for law enforcement.[70] In 1987, German BBS sysops received unexpected visits from post office officials searching for modems and acoustic couplers.[71]

The privacy of those digital publics and counter publics were first and foremost protected by the German constitution. The German Basic Law from 1949 lists "inviolability of mail, post, and telecommunications" as a fundamental right, only restricted pursuant to a law, but also imparted a duty on the state to provide essential communication services to the country.[72] This means that while Germany allows for the interception of communication following procedural requirements, Germany would not engage in "consent surveillance"—all parties must consent to authorize access to communications.[73]

Like in other countries, two sets of procedures exist for national security and criminal prosecution. Article 10 of the Basic Law was amended in 1968, establishing that infringements on the right to communication privacy could be kept secret to protect national security, and recourse would be removed from the courts and instead reviewed by a special body to maintain secrecy. This Basic Law amendment was accompanied by what is known as the G10 Act of 1968. The amendment and law were passed to enable West Germany to assume responsibility for its own security, after ceding all surveillance power to Western Allies for two decades after World War II. Surveillance for both prosecutorial purposes and national security included restraints like requiring investigators to provide written form and scope of the surveillance, limiting the time period of surveillance, and obligating officers to discontinue it as soon as it was no longer necessary for the stated purpose. Prosecutorial and national security procedures differed in that the former included the standard for reasonable suspicion of a punishable crime to authorize surveillance, and the latter only required an administrative order (as opposed to a judicial order) but was overseen by a G10 Commission with additional reporting requirements.[74]

In 1978, weeks before James Malone went on trial for stolen property before the British courts, the European Court of Human Rights in Strasbourg interpreted article 8 of the European Convention on Human Rights as it applied to the German surveillance law in *Klass v. Germany*.[75] The plaintiffs argued that the lack of judicial pre-approval in national security instances and the lack of universal post-surveillance notification to those who were wiretapped violated article 8 of the ECHR. In 1970, the German Federal Constitutional Court upheld the basics of G10, including the executive warrant system, but required notification when it would not endanger the monitoring goals.[76] Subsequently, the G10 Commission also began to oversee decisions to withhold notification. By the time the ECtHR took up the statute, it was able to uphold the law. The court began with a passage that is now, without fail, cited in all post-*Klass* cases. It explains the ECtHR "must be satisfied that, whatever system of surveillance is adopted there exist adequate and effective guarantees against abuse."[77] Ideally, those guarantees include judicial control, and the court found the German system satisfactory and agreed that notification is not always possible, but should not be prohibited entirely.

The provisions that govern nongovernmental parties' access to communications are found in chapter 15 of the German criminal code, wherein an unauthorized person (*wer unbefugt*, translated to "whoever, without being authorized to do so") is prohibited from unlawfully recording privately spoken words to another, sharing such a recording with third parties, and using a device to unlawfully listen in on private spoken words without authorization from the parties involved.[78] This criminal wiretap law was soon complemented by specific computer crime legislation as the CCC's reputation morphed.

The Chaos Computer Club unsettled users' sense that the BTX system was secure, but its early antics were celebrated, not punished. In 1984, the CCC exploited a memory glitch in the way BTX stored information and transferred 135,000 Deutschmark through the CCC donation page. They quickly returned the money and then pressed charges against themselves, making it all the more ridiculous to prosecute the group for pointing out the flaws in BTX. The bank thanked them.[79] But authorities were beginning to lament a lack of computer crime legislation. A few months before the BTX exploit, experts expressed an urgent need for legislation to address the "American" trend of computer sabotage and "time theft" (stealing time on a mainframe) that had arrived in Germany. This consultation was tacked onto an effort to update white-collar crime laws. The CCC and German hacker culture were not central to the experts who updated the *Strafgesetzbuch* (StGB) criminal code, but these changes would apply to them. Section 263a StGB was added to define computer fraud; section 303a was added to address alteration of data; section 303b punishes computer sabotage. The statute vaguely applied to a "non-authorized person" who entered incorrect information, manipulated the processes of a computer program, or changed, blocked, or deleted accurate data from being processed.[80]

The cinematic American hacking trend had in fact arrived in Germany, most famously with the Hanover Hackers, as the press dubbed them. The Hanover Hackers were a group with ties to the CCC who hacked into an American astrophysics lab's supercomputer and were caught accessing military computers in 1987 for the Russian KGB. Cliff Stoll's *The Cuckoo's Egg* is a first-person account of the hack. In it, the lab's system manager and resident sleuth (Stoll) have the following exchange:

"The real problem is in German law," Steve said. "I don't think they recognize hacking as a crime."

"You're kidding, of course."

"No," he said, "a lot of countries have outdated laws."[81]

Germany was "up to date" though, having passed a suite of computer crime laws in 1986.[82] These included section 202a, which referred to "data espionage" and addressed access to data without authorization, as well as data theft.[83] As the CCC fought to keep users from becoming Lusers, German law made sure users were authorized and hackers were not.

LUSER SCREEN NAMES: ONLINE SERVICES IN THE US

By 1986, Minitel was handling 47 million calls a month, and videotex services in the United States were handling only around half a million.[84] In an earlier Federal Trade Commission report, a committee analyzed Prestel, Telidon, and Minitel, explaining, "We have chosen these particular systems for two reasons: they reflect the major design distinctions among such systems and they appear to be the systems most likely to have substantial penetration in the U.S. in the near future."[85] Indeed, American partners were testing each project in different regions and appealing to the Federal Communications Commission. None were ever considered successful. Interest in US videotext systems in the 1980s remained high because of videotex's technological capacity to become a marketing tool. One executive explained, "Videotex is a magnificently precise measurement tool . . . With it you can track how someone uses the service, what he reads, what he does. That is what makes it so attractive to big business."[86] This capacity was not seen by all as an opportunity. Because the videotex systems were centralized and designed for the home, there was concern that system operators would use the central computer to create a "home profile."[87] One lawyer warned at the time, "[Videotex] systems operators will be collecting massive amounts of personal data from subscribers. Whether the subscriber is ordering goods and services, answering inquiries, retrieving information, or utilizing the security services, he will be conveying his interests, choices, and views to the central computer."[88] But the surveillance tool of choice in the US would not be videotex. The American computing market and industry presented opportunities for a different type of network experience that came to be called "online services," where early forms of identity practices and advertising experiments played out.

Commercial online services in the US were innovative not so much for the technology that they created but for the access and ease with which they deployed the technology. Most commercial online services relied on a similar infrastructure as the hobbyist bulletin board systems, wherein a modem and a telephone line were used by callers to access a remote host server that offered interactive services. However, instead of a caller having to visit multiple niche bulletin boards to explore a smorgasbord of content and interact with different users, commercial services endeavored to give users everything they wanted in one place. They moved away from "dumb terminals" (a simple terminal that depends on a host computer for processing power) and harnessed the power of the new personal computer models being released by Commodore and Apple.

America Online (AOL) was one such company. AOL is an iconic US company and in many ways is responsible for getting Americans on computer networks. With the goal of making it cheap and accessible to connect to its online services, AOL evolved from a Commodore-specific network called Quantum Computer Services to become one of the world's most prolific online service providers. AOL became the online service for the everyday person, and its content reflected that, as it provided games, news, weather, and community forums. Stephen Case, the CEO credited with making AOL a behemoth among online service providers, explained, "Every day, I wake up and say, how can we make America Online more interesting, more useful, more fun, more affordable, so that it will attract a broader audience?"[89] In 1994, AOL provided access to Usenet, locatable if users selected the Internet Connection button (among Today's News, News Stand, Clubs & Interests, Education, Sports, Personal Finance, Travel, and Marketplace). Usenet is a forum or message board system similar to a BBS but does not use a central service or designated administrator (though connections were made between the two), and users read and post to topic categories called newsgroups. Usenet users noticed the flood of new clueless entrants in 1993 when internet service providers began connecting subscribers to Usenet, but when AOL offered widely accessible connection in September of that year, the influx became known disparagingly as the Eternal September.[90]

From its beginning, AOL was focused on user communication and digital presence rather than on just becoming a means to distribute information, and users flocked to the service. Because of AOL's popularity, the service's practice of requiring a "screen name" to log in online became the norm for all online communities. AOL allowed the usage of five screen names per account, each with its own password and email box.[91] Online services came to be called "walled gardens" because they were confined and curated by platform companies like AOL that would shape the rules and culture within them.[92] Nonetheless, compared to BBSes, AOL's rules left matters of user identification up to users.

While the main privacy concerns on AOL were about managing information seen by other peer users, Prodigy users were concerned with structural corporate invasions of privacy. Prodigy officially launched on September 6, 1990, as a partnership between IBM, AT&T, and Sears. Although it launched after popular online services like CompuServe and The Source (Source Telecomputing Corporation), Prodigy claimed to be the "first consumer online service."[93] Unlike the other online services, Prodigy was a videotex system that used the X.25 dialup standard to connect to IBM mainframes and charged a flat monthly usage rate. To help keep costs down, Prodigy used a system that it called "points of presence" or PoP, in which relevant content was cached in computers distributed to 120 PoP sites. The cached content would have to travel shorter distances to reach the user, so each unique user did not have content sent from a central mainframe. Additionally, Prodigy could keep costs low by carrying advertising. A large portion of advertising came from merchants on Prodigy's home shopping network and took up about a quarter of a 24-line screen of videotex. These sections of the screen were sold to advertisers on a per view basis and came with consumer data—a fact that did not go unnoticed by the small portion of the public who were paying attention to online services. "Choosing among the network's hundreds of features and by purchasing items through the home shopping network, subscribers provide all sorts of demographic information to Prodigy and its advertisers," reported the *Boston Globe*.[94] The model triggered privacy concerns among some user groups, including panic over a type of file-writing that Prodigy did on users' computers. Prodigy, like other commercial online services of the time, wrote temporary data in files called "stage.dat" and "cache.dat" that it stored on a user's computer. Eventually someone noticed that data from other applications like word processors on the computer ended up in the Prodigy-created folders. As reported in the *Wall Street Journal*, the stray bits of data were written into Prodigy files as part of a "fluke" in IBM-compatible operating systems. Prodigy sent out free software that would ensure the deletion of any data in stage.dat or cache.dat.[95] Despite this intervention, because Prodigy was already openly monitoring its users' activity for advertising purposes, the idea that the company might install spyware on a user's computer gained momentum. In an interview with the *Los Angeles Times*, a computer analyst who initially thought the spyware scare was "a lot of horse hockey" admitted that she canceled her Prodigy subscription after looking into her own files.[96]

Online services operated with relative legal certainty, enjoying the confidence provided by a slew of computer privacy laws that passed in the 1980s to support precisely those computer communication industries. In the US, the Fourth Amendment provides the constitutional protection from governmental (though not private sector) invasions of privacy, ensuring rights against government interception of communications. Case law delineates between the content of the communication (the letter), which gets a great deal of protection, and the non-contents (the envelope), which gets virtually none. The distinction is based on an expectation of privacy associated with what is necessary to transfer the communication versus the substance of the communication. Once information is shared, individuals waive their right to privacy because the information has been exposed to others. Cracks in this "third-party doctrine" have only very recently begun to show.[97] Paul Schwartz captures this aspect of US surveillance law succinctly: "In the United States, one bears the risk that any party to whom one reveals her affairs will share this information with the government."[98]

Even before the Fourth Amendment played a significant role, the Postal Service Act of 1792 had already distinguished and shaped American privacy expectations and practices regarding communications. American colonies previously relied on the British Royal Mail, a classic example of the state-owned system of control and surveillance of the time. American rebels attacked British postal officers, stole mail bags, set up alternative routes, and took over local offices, eventually crippling the British post completely and establishing their own Constitutional Post. The first American lawmakers rejected the longstanding presumption that governments had the power to open letters and banned the practice of opening letters unless authorized by a warrant.[99] Postal officers who opened letters could be severely punished—with death, if the letter contained money. Designed to engender trust in the postal service, the law did just that, largely achieving its civic and commercial goals. But there were also practical issues with opening letters. Compared to the highly centralized and professionalized service in Europe, the US post was decentralized and local, which meant it was just not possible to inject a smooth surveillance apparatus that would not interrupt service.

Uniquely provided by corporate actors instead of the US government (despite initial funding from the post office and efforts by postal leaders), telegraph services required technical mediation that complicated the roles of communicators and notions of privacy. Typically, a customer would write out their message on a piece of paper and then hand it to an operator, who would convert it and transmit it to another operator, who would do the same, interpreting and translating the received message, writing it on a piece of paper for the recipient. Decorum and ethics, as opposed to policies and laws, were promoted to ensure trust in the system. In an 1860 essay "Secrecy of Telegraphic Communication," one telegraph engineer insisted there existed a "high sense of honor which every operator feels upon this point," and New York and Pennsylvania had already passed laws making it a misdemeanor for operators to divulge messages without sender or recipient consent.[100] The Revolutionary War sentiment of distrusting a surveillant post did not last through the Civil War. President Abraham Lincoln, with the support of Congress, seized control of telegraph lines and monitored messages. After the war, telegraph companies, namely Western Union, had to walk a fine line. They needed to transmit confidential messages for their customers (especially those members of the press who condemned those who handed over messages to the government) but were also worried about government takeover.[101] When a Western Union operator who had transmitted messages relevant to leaked details of the Treaty of Washington was called before Congress, he refused to testify. He, like many operators before him, explained he could not divulge the content of communication without violating his professional duties. He was celebrated by the press, which used privacy as a justification for maintaining private ownership of the system. By the late 1870s, the assertion of privileged communication transformed into objections derived from the Fourth Amendment protections from government invasion, due in no small part to the Constitutional call made by the prolific Michigan Supreme Court Chief Justice Thomas M. Cooley in his article "Inviolability of the Telegraphic Correspondence"—a call taken up by Warren and Brandeis in their seminal 1890 article "The Right to Privacy."[102]

When Brandeis later wrote his famous dissenting opinion as a Supreme Court justice in the 1928 case *Olmstead v. United States* challenging the constitutionality of wiretapping a phone without a warrant, he maintained his innovative privacy position. Former police officer turned "king of bootleggers" Roy Olmstead disputed the use of wiretap evidence to support his conviction, but the majority of the Supreme Court found no violation of the Fourth Amendment. Chief Justice William Howard Taft argued that it "does not forbid what was done here. There was no searching. There was no seizure. The evidence was secured by the use of the sense of hearing and that only."[103] Telephone use didn't require sharing the contents of communication with an operator, though it required help from an operator who knew who was calling whom. But the social terms of surveillance shifted significantly in the years that followed *Olmstead*, which included a world war and the rise of "law and order" politics.[104] The legislation that updated communication privacy for telephones was passed in response to two 1967 Supreme Court decisions: *Berger v. United States*, which determined that a New York state law did not require precise or high enough administrative hurdles to meet Fourth Amendment standards, and *Katz v. United States*, which found that wiretapping a public telephone booth was an unconstitutional invasion of privacy despite the absence of police trespass.[105] Title III of the Omnibus Crime Control and Safe Streets Act of 1968 provided detailed specifications for wiretapping and an explicit exception when one party to the communication gave "prior consent" to interception.[106]

In 1986, Congress notably and explicitly added the user to American privacy law when it passed the Electronic Communications Privacy Act (ECPA).[107] ECPA extended protection from interception beyond voice to electronic communications, and section 2510(13) defined the user for the first time. The user was and is, "Any person or entity who uses an electronic communication service and is duly authorized by the provider of such service to engage in such use." ECPA also changed the use of "any communication common carrier" to "a provider of wire or electronic communication service."[108] The first hearing included in the legislative history is from 1975, and it considered bills that would have limited government use of electronic eavesdropping, which were intended to address the continued public outrage. By the 1984 hearings, computers were front and center. Senator Patrick Leahy opened the subcommittee hearing by stating:

We have phones ringing all over the country, answer it, and it is not voices you hear but dots and zeros and blips and beeps that come out of it. And that is the information that is going in digit form and this is everything from interbank orders to private, electronic mail hookups. It is nothing remarkable, but it is remarkable that none of these transmissions are protected from illegal wiretap, because our primary law passed back in 1968 covered only voice transmission; it failed to cover nonaural acquisitions of communications of which computer to computer transmissions are a good example.[109]

Phone lines were being used to transmit a lot more than voices. The digital communication sent by users that was transforming phone lines into computer networks lacked the same legal privacy protection.

A year later, in a Senate hearing to discuss the ECPA bill, Congressman Carlos Moorhead of California testified in support of protecting "users of any of the new electronic communication systems." He had recently met with members of the Cellular Telecommunications Industry Association who were concerned that the 1968 Wiretap Act did not protect their "users."[110] Testimony from a panel of witnesses reinforced this idea of the user as working at the convergence of electronic communications. Philip M. Walker, the vice chair of the Electronic Mail Association, which started in 1983 and had over sixty members, spent a lot of time explaining the basics and benefits of electronic mail. He projected that, while the computer-based messaging industry was valued at about $250 million in the mid-1980s, it would grow to $2 to $3 billion in the 1990s, and that electronic mail would become a regular part of the American workplace, but that would require specific privacy protections for "users."[111] The "Authorized User" in the final version of ECPA was an edit suggested by the Association of Data Processing Service Organizations (ADAPSO). The trade association explained that substituting "a user" with "an authorized user" was necessary to prevent "unauthorized users, who are nonetheless 'users,' from 'authorizing' and thus legalizing improper access by another."[112] The group went on to suggest spelling this distinction out in the definition section. In the House of Representatives, where ECPA was largely developed by the Subcommittee on Courts, Civil Liberties, and the Administration of Justice, The Source promoted the exact same revisions. The Source was one of the earliest online service providers for the general public and one of the only to provide testimony that mentions "computer users" in the hearings.[113]

ECPA broke the previous law into three parts: the Wiretap Act in Title I, the Stored Communications Act in Title II, and the unnamed Title III, which addresses pen registers. The Wiretap Act applied and continues to apply to communication in transit, while the Stored Communications Act governs when the information is being stored. As the previous two chapters detail, this is similar to the distinction Europeans draw between privacy and data protection. The comparison causes confusion because Americans have packaged them both into the communication privacy law, while Europeans have distinct legal frameworks with robust rights and obligations under the data protection regime.

The Wiretap Act made it illegal for anyone to intentionally intercept any wire, oral, or electronic communication. The exceptions to this blanket provision include law enforcement with a special warrant, electronic service providers acting in the normal course of business necessary to supply services (referring to switchboard operators and technicians), and parties to the communication who have provided consent. Wiretaps require some of the highest administrative demands for law enforcement in US law, including judicial approval and heightened scrutiny. These demands and scrutiny don't matter if consent is provided from "the originator or any addressee or intended recipient of such communication."[114]

The Stored Communications Act contained a similar exception for consent provided by an authorized user. Access "to a communication of or intended for that user" is simply not prohibited if authorized by an authorized user of the service.[115] The SCA prohibited intentionally accessing an electronic communication service without authorization or in a way that exceeds authorization, except when user consent has been given or the government has met its relatively low administrative burdens (which vary depending on how old the communication is and whether it has been opened). Additionally, it contains detailed restrictions on when electronic communication service providers can divulge content of communication, including when governmental entities meet their administrative burdens or "with the lawful consent of the originator or an addressee or intended recipient of such communication."[116] The SCA also states that electronic communication service providers may divulge non-content information to "any person other than a governmental entity."[117]

Access by outsiders was also criminalized by the passage of the Computer Fraud and Abuse Act (CFAA) in 1986. The personal computer was having a moment in the US during the early 1980s, much more so than in Europe, where France proved the exception by providing Minitel terminals in the millions. IBM released its first PC in 1981 and the Commodore 64 followed a year later. Still, when Congress began considering computer crime legislation in 1983, there were only 3.5 million personal computers in use and only about 200,000 had communications capability to access remote computers, with a projection of seven million by 1986, two million of which would have remote computer communication functions, and ninety million with unknown functionality by 1990.[118] Like other computer crime statutes, the CFAA criminalizes "whoever" for accessing a computer without authorization or exceeding authorization under various circumstances.[119] Who was the "whoever" Congress had in mind? The 414s.[120] The CFAA authors fretted over the very real Milwaukee teenagers living in the 414 area code who met through an IBM-sponsored Explorer Scout troop and were inspired by—you guessed it—*WarGames*. The 414s managed to gain access to numerous high-profile systems, including the Memorial Sloan Kettering Cancer Center and the Los Alamos National Laboratory, a site for nuclear weapons research.[121] After appearing on the Today Show, CNN Crossfire, Phil Donohue, and the cover of *Newsweek*, Neal Patrick, the only minor of the 414s, became a celebrity and bantered about whether computer trespass was "prank or sabotage."[122] The first computer Patrick touched was the mainframe set up as part of the state's educational network, one of many education networks developed in different US localities that promoted access and use for all.[123] When Patrick testified before Congress in 1983, he was asked, "At what point did you first question the ethical propriety of what you were doing?" He responded, "Once the FBI knocked at my door," to booming laughter.[124] The ethical acumen of other computer users was at issue as well. A 1984 report by the American Bar Association's task force on computer crime found that half of the responding companies and government agencies reported being victims of some form of computer crime, resulting in as much as $73 million in damages. This new breed of criminal shared passwords, posted credit card numbers, and exchanged "program-devouring programs" on "pirate bulletin boards," which were cast similarly to the German account when authorities cracked down on BBSes to suss out CCC members.[125] The vice president of CompuServe, described in 1984 as a "remote computing service organization" with more than 110,000 subscribers, impressed upon Congress that "the costs that unauthorized users inflict occur in several different forms" and that the underground marketplace for trading in unauthorized use on bulletin boards was robust.[126] The space and activities were described as existing in a legal gray zone. Because there were no computer crime laws, the 414s could only be charged with making harassing phone calls. The subcommittees intended to separate computer use into black and white. Through the CFAA, they tried to separate hackers from authorized users (lusers).

EPRIVACY USERS: EURONET IN THE SINGLE MARKET

Although the centralized network of the late 1980s and early 1990s online services was not the future of networking, the services eventually provided access to Usenet and the World Wide Web, which was where the future lay. In 1989, CompuServe allowed email exchanges with internet-based email addresses. In 1994, Prodigy was the first to provide full access to the World Wide Web. American users moved onto the web with unique protections for platforms and consent-based exceptions to wiretap laws. Network competitions and entanglements continued across the Atlantic. Because of the infrastructural, cultural, and regulatory barriers, AOL faced stiff competition and a complex marketplace when it partnered with Bertelsmann Group in 1995 to launch its services in Germany, France, and the UK.[127] As did Europe Online, a continental commercial online service backed by AT&T and a number of publishers and banks, which was introduced in English, French, and German with plans for expansion.[128] Europe Online was an express, politically supported response to the Americanization of cyberspace. A representative of Europe Online explained that behavioral and demographic data would not be given to advertisers: "The feeling in Europe is that an online service built for and by Europeans is needed. They need a service that is done by people with a deep understanding of the European situation."[129] Europe Online was only one such effort.

While there had been a mix of interest from traditional providers in packet-switching networks, the European telecom monopolies agreed that they would pursue their own standard in an effort to diminish American power over the global computer market. As described above, work toward digital networks related to the internet began in 1970 and was organized at the national level, and while they used packet switching, the networks were limited and not interconnected. The focus of the European Council in the late 1970s and early 1980s was on cooperation and harmonization of these networks through Euronet to support economic growth across the European Community.

Euronet was a Pan-European internet intended to become a vast global public network that would end dependence on US information retrieval.[130] Formally launched with the passage of a resolution by the Council of Europe, Euronet began as plans to coordinate an online information network for sharing science and technology documents. The first stage of the plan was to create the physical connections of the nine PTTs. Using the X.25 standard, the Council-created Committee for Information and Documentation on Science and Technology installed switching nodes in major cities. The second and third stages expanded to the general public and beyond science information. Euronet was open to the public but a closed system. The PTTs maintained control over how and what data traveled over the network. It cost over $20 million at the time.[131] In a special session of the European Parliament on February 13, 1980, Euronet connected all (then nine) Member States to Direct Information Access Network for Europe (DIANE), a collection of hundreds of databases. Euronet offered cheaper per search costs and no long-distance charges—there were no plans for a transatlantic link.[132] In its coverage of the launch, however, *The Economist* noted, "The hope is that it will wean European companies and universities away from their present dependence on American data retrieval. That seems unlikely: Euronet's technology may be nearly obsolete."[133]

The EU was not interested in supporting the internet or any other non-European network standard, and instead championed its X.25 standard. The X.25 networks that made up Euronet slowly lost significance over the course of the 1990s, but the EU market that the standard was intended to support remains the largest in the world today. With the fragmented nature of European markets identified as a weakness in relation to US players, national deregulation, institutional privatization, and regulatory unification occurred across European technology markets under the focused agenda of the European Commission.[134] The Single European Act of 1987 dramatically revised the EEC (soon to become the EU) for the first time and sought to establish a single internal EU market.

In pursuit of a single EU market over the course of the 1990s, the EU passed a series of directives that required EU Member States to phase out the state communications monopolies and welcome free market forces to the sector.[135] Full liberalization of the telecom market in the EU was to be complete by 1998. With this shift away from a government monopoly to the private sector, existing regulatory power diminished. Part of the deregulatory agenda included a new regulatory package, notably the ePrivacy Directive. Although the EU Telecommunications Directive was adopted in 1997, it drew heavily from the Council of Europe's recommendations on the same telecommunication subjects, focusing on the telephone.[136] Despite that, the user was mentioned throughout and defined alongside the "subscriber." Implementation of the directive was stalled by the European Commission's review of regulatory approaches to electronic communication in 1999. The review resulted in a call to replace the Telecommunications Directive, which focused on traditional telephony services, with a "technology neutral" approach that also reached electronic networks and services.[137] This negotiation produced the 2002 ePrivacy Directive, which set out to protect "privacy for users of publicly available electronic communications services, regardless of the technologies used."[138] With its adoption, each EU Member State was required to pass domestic laws to protect the communication privacy of the user. Specifically, Member States had to pass laws that ensure the confidentiality of communications on public networks and publicly available electronic communication services in particular. These domestic laws had to "prohibit listening, tapping, storage or other kinds of interception or surveillance of communications and the related traffic data by persons other than users, without the consent of the users concerned, except when legally authorised to do so in accordance with Article 15(1)," in which proportionality and balance with public safety, law enforcement, and national security are demanded.

The user was the central figure protected by the ePrivacy Directive and the first term defined, using the same language as the 1997 directive. Article 2(a) states, "'User' means any natural person using a publicly available electronic communication service, for private or business purposes, without necessarily having subscribed to this service."[139] All Europeans on electronic networks were users by the new millennium, and user consent was a powerful form of authorization. Beyond authorization through a bureaucratic process that met EU human rights standards, only consent of users could justify intercepting communication over European computer networks.

Of course, national and regional efforts to create computer networks—some of which were funded by the European Commission and restricted access to European Community members or pressured participation to promote European market players—were eventually supplanted by a global network of researchers and computer industry participants using TCP/IP protocols dominated by American private actors intent on cooperation without regulation.[140] Over that ongoing and arduous process, during which numerous discontinuities emerged on forgotten and obsolete networks, the user was legally constructed through the efforts of many countries.[141] The user was consistently held to be computer savvy enough to authorize access to computer information but not savvy enough to find themself on the wrong side of hacker laws. The user was a far cry from the data subject, who existed in a compromised state that required giving up data and who was given powers to control computers in order to limit the power of those employing computers to control data subjects. The data subject was created for computers. The user, on the other hand, was a part of existing communications practices that relied on parties to consent to interception, access, and use of communication content and details. The user emerged to help refine aspects of existing communication and criminal law. The user is also markedly different from the third computer character relevant to cookies, the privacy consumer, who was made in the US to promote choice on the commercial web built on the legacy of these networks.

4 MAINTAINING STATE FOR THE PRIVACY CONSUMER

The third and final computer character who consents to cookies is the privacy consumer who was made up for the commercial web. Although the privacy consumer has important ties to the 1960s (the decade that introduced the data subject and the user), they were not born until the late 1990s. In the decades prior, a handful of creative innovators mainly at universities played with ways to arrange, connect, and engage information on computers. Tim Berners-Lee joined that group as he and the team at the Swiss European Organization for Nuclear Research (CERN) launched the World Wide Web at the end of the 1980s. Later, through negotiation and experimentation in pursuit of a commercial web, developers found ways to address the web's "stateless" design. When a computer requested information over the web, the server did not store or memorize information about the requesting computer (the "state" of the system), so each interaction was new. The group debated what exactly should be remembered, how, for whom, and to what ends—and whether consent was required to set up any such memory system. Over a short few years, American entrepreneurs, developers, and policymakers transformed the protocols created in Switzerland for sharing scientific documents into the commercial web, ushering a new computer character into prominence: the privacy consumer.

A new set of advertisers fought for a vision of the web where advertising served as infrastructure to support a decentralized web, while at the same time the privacy consumer emerged to make consumer choices about sharing information with an array of sites and services. In a historic moment of powerful consumerism, the privacy consumer was bestowed with the capacity to choose, and the act of choosing was understood as an act of privacy protection no matter the choice. However, the default setting for technical standards and legal policies would need to support the economic viability of the web, or so the argument went. Stateside, through the efforts of the Clinton administration and online advertising associations, the privacy consumer for the web was situated within long-standing debates about junk mail. Mailing physical advertisements directly to potential customers with more precision was the business of direct mail marketers, a small group that began to form in the 1960s and 1970s before finding success and growing in the 1980s—and with it the amount of junk mail.

By the 1970s, in response to complaints about junk mail, direct mail marketers had developed an opt-out system that passed scrutiny in early US privacy reports, wherein consumers could opt out of receiving junk mail by filling out a form and mailing it back. When the Clinton administration situated the issue within the limits and logics of the Federal Trade Commission (FTC) and the Department of Commerce (DOC), policymakers further cultivated the system within the administration's aggressive efforts toward commercial innovation. The online advertising apparatus moved abroad quickly, bringing with it the rhetoric of advertising infrastructure and the role of the privacy consumer, but the practicalities of the market proved challenging. Political resistance came later, as detailed in the next chapter. Over a period of just a few years in the late 1990s, online data became consumer data, internet privacy become consumer privacy, and users became privacy consumers.

THE STATELESS WEB

The web that would produce the powerful privacy consumer came from activity around hypertext. People around the world working on or with hypertext on various computers and networks in the 1980s sought to retrieve and display information in new impactful ways. In 1987, Belgian information engineer Robert Cailliau became the head of Office Computing Systems in the Data Handling division, where he promoted the idea of a hypertext documentation system for CERN. By then, CERN offices were networked with AppleTalk and AppleShare, but in fact, CERN had itself developed several similar prior innovations in the mid-1970s. Back then, Cailliau was a new CERN employee working on the controller system for the Proton Synchrotron particle accelerator. CERN had its own network code that could be moved from one machine to another and its own byte interpreter (essentially Java and the internet), as well as its own interfaces.[1] Cailliau explains, "The whole thing was done on Norsk Data computers. They were a small Norwegian company—the last independent computer company in Europe, I think—and after that nothing existed except from the U.S., and of course from the U.S. perspective, if it isn't done in the U.S. it doesn't exist in computing, right?"[2]

Cailliau knew about hypertext but didn't know about the internet. Tim Berners-Lee, a young British engineer who worked in another division, knew about both. In 1989, Berners-Lee wrote up and proposed the World Wide Web for his boss, who connected Cailliau and Berners-Lee.[3] By the end of 1990, the first web page was up and running. It was built on elements Berners-Lee had also written during that year: HTML (Hypertext Markup Language), HTTP (Hypertext Transfer Protocol), URI (Uniform Resource Identifier), and the first web browser (WorldWideWeb.app).[4] However, Berners-Lee built his WWW for Steve Jobs's NeXT platform, and the team did not know how to move the WWW application to the X Window System, which was in far greater use among science institutions. Nicola Pellow, a British undergraduate math intern who joined the WWW Project in November 1990, created a less flashy Line Mode Browser that would run on other operating systems using the command-line interface.[5] Her version was made available on CERN computers in March 1991, and to internet newsgroups also seeking collaborators in August 1991.[6] All of it was nonproprietary, but the Line Mode Browser had no editing capabilities.[7] Cailliau explained, "It gave the Internet hackers immediate access, but only from the point of view of the passive browser . . . It was quite depressing, having to step back and say, 'I have to port this to a PC?'"[8]

As the face of CERN's WWW, Berners-Lee had entered the world of hypertext, a community that used computers to rethink media. Vannevar Bush's theoretical Memex machine is often cited as the beginning of the hypertext story. In his 1945 article "As We May Think," Bush envisioned a machine for self-reflection that would store, organize, and create associative trails among information recorded on coded microfilm frames.[9] Inspired by this machine and the idea of computers as more than institutional calculating machines for military and scientific mathematics, Doug Engelbart displayed hypertext functionality in his Mother of All Demos in 1968.[10]

Bush's Memex also inspired Ted Nelson, who coined the term "hypertext" in a paper presented at the Complex Information Processing track of the Twentieth National ACM Conference in 1965.[11] The article received little attention.[12] Nelson had no technical training or skills, but continued to work on the idea of hypertext media as a personal project while sporadically employed after college. As one writer put it, "For Nelson, paper was the enemy."[13] This was a problem, because writing at the time meant paper, and computers meant numbers.[14] Nelson found his way to computer science professor Andries van Dam, who was working on an early document Hypertext Editing System (HES) at Brown University, but when van Dam shopped the program around to places like the *New York Times*, he was told it had too many buttons and there was no way journalists would sit at a computer.[15] Van Dam was in the audience for Engelbart's Mother of All Demos and afterward he went back to Brown to create the File Retrieval and Editing SyStem (FRESS), which was popular due to its undo and redo functions. FRESS was made not for physics or engineering or lab technicians, but for writers. Although van Dam constantly had to fight for computing time on the university mainframe, he was able to use the system to teach poetry in 1975 and 1976.[16] Engelbart reflected, "The idea was wacky even in the seventies, when we had it working—real hypermedia, real groupware working."[17]

When the Apple HyperCard introduced the public to hypertext in 1987, it was a move beyond Nelson's paper enemy. However, it remained a stand-alone workstation-style database system, not the global "intertwingularity" he preached.[18] British mathematician turned computer scientist Wendy Hall became interested in the possibility of hypertext after seeing the BBC Domesday Project.[19] The project was a multimedia presentation of an eleventh-century English census that required a BBC Master AIV interactive video controller, a laser disc player, and a display monitor.[20] Hall joined the newly formed computer science group at the University of Southampton in 1984, and the team created Microcosm in 1988. Microcosm represented an alternative vision of links—it was social, multidimensional, and evolutionary. Instead of embedding links into a document, image, or video itself, Microcosm maintained a "linkbase" of automatically generated, updated, and maintained links based on the metadata of the content. It was a fundamentally different way to think about associative trails through hypertext.[21] In 1991, when Berners-Lee and Cailliau showed up at the fifth Hypertext conference, their WWW seemed rudimentary because of its unidirectional links, and nichey because it required an internet connection.[22] At the time, Hall found the name pretentious, but by 1993, half of the papers presented at the conference were WWW papers.[23] By 1994, Berners-Lee, who had moved to MIT and started the World Wide Web Consortium (W3C) to help technologically organize and further the web, gave the keynote address at the Hypertext conference.[24]

Lou Montulli, a student in academic computer services at the University of Kansas, was working on essentially the same project around the same time. Montulli worked with staffers Michael Grobe and Charles Rezac to build a text-only hypertext browser called Lynx for campus documents housed on Gopher servers. Gopher was created at the University of Minnesota in 1990 by a team sick of attending meetings about what to do with the Minnesota network. But when they proposed a campus information system that operated on the trendy new PC instead of the mainframe, the rest of the committee revolted.[25] Nonetheless, they designed Gopher for non-technical users. The main menu was a list of folders that included information about Gopher, games, mailing lists, phone books, and campus information. Lynx was a very popular Gopher browser, but in 1992 Montulli began integrating the library from CERN and adding HTTP support. In July 1993, Grobe and Montulli attended "The Wizards Workshop," which was a precursor to the 1994 "Woodstock of the Web," the first of the World Wide Web conferences that still run today.[26] Grobe explains that they "invented *a* Web . . . [r]ather than *the* Web."[27] Other browsers were built at universities, such as MIDAS at the Stanford Linear Accelerator Center, Viola at the eXperimental Computing Facility at the University of California, Berkeley, and Erwise at Helsinki University of Technology.[28]

And Mosaic in the National Center for Supercomputing Applications (NCSA) at the University of Illinois—the university browser that would become the commercial browser credited with driving people to the web. The NCSA was a government-funded effort that was quickly becoming passé in the face of smaller, more powerful computers. But the University of Illinois served as the original backbone of the National Science Foundation Network (NSFNET) and had perhaps the fastest internet connection in the world. Marc Andreessen worked at the NCSA as an hourly computer science student employee. In 1992, when there were twenty-six WWW servers in the world, one was at the NCSA.[29] Andreessen began working with Eric

Bina and Colleen Bushell to create a browser that someone could use without earning a computer science degree. Bina did much of the coding and Bushell contributed the interface design, while Andreessen took the lead on envisioning and managing the project and extra features. Andreessen posted X-Mosaic, which ran on the Unix X Window system, to the NCSA server in January 1993, and Berners-Lee forwarded it to newsgroups. Other NCSA programmers signed on to create versions for PC and Mac (just called Mosaic), as well as developer and publishing tools. Mosaic's graphical interface was a showstopper. Bushell had trained as a graduate student under the computer visualization pioneer Donna Cox. The Mosaic browser was more intuitive and easier on the eyes than alternatives, and Bushell injected the aesthetics with thoughtful, exciting innovations like an animated browser logo that indicated data transfers (a loading icon).[30] It's considered one of the earliest developments of its kind. Sitting on the porch of the old house she and her husband were renovating, Bushell, Bina, and Andreessen celebrated with a margarita. Mosaic had taken off. Within months, the Mosaic browser had been downloaded hundreds of thousands of times, and by the end of 1993, there were 500 known web servers, which together accounted for 1 percent of all internet traffic.[31]

Upon seeing Mosaic, Cailliau said, "It was okay, but it was a single window thing. A single window, non-editing thing that got its popularity from two aspects: it was much easier to install than any of the other, better, X-Window based browsers that went before it because it came as one big blob for Unix machines. Its second characteristic which was very attractive was that it was close to what people knew: it put the images in line."[32] Using Gopher and other WWW browsers, images could only be viewed by clicking on a link that would prompt them to download in a new window. Andreessen created the "img" tag so that an image could be displayed in a web page and downloaded while the text of the page was available.[33] When Montulli told the others on the www-talk mailing list about creating a new HTML element, Berners-Lee tried to get him to just use the existing "anchor" tag so users could set their own preferences for handling images. But Berners-Lee's customizable, editable browser experience was not the focus for the Mosaic team.[34]

Mosaic's single window impressed Jim Clark. Clark is a famous figure in Silicon Valley, having started Silicon Graphics (SGI) in 1982 after holding positions at Xerox Park and Stanford. SGI built terminals and workstations that could render 3D graphics, and it was one of the most successful companies in the Valley throughout the 1980s. Clark slowly lost control of his company as it doubled down (against his wishes) on high-end specialized workstations instead of pivoting toward cheaper and more commercial machines. He left in 1994, armed with many lessons and ready to start a new company, but he wasn't sure what direction to go and couldn't poach any of the SGI engineers. A friend suggested Andreessen and showed him Mosaic. Clark immediately reached out to Andreessen. After kicking around a bunch of ideas, the two incorporated Mosaic Communications Corporation in April 1994. Clark and Andreessen recruited the Illinois team, and Montulli flew in from Kansas. Despite the fact that the interface design was one of Mosaic's most important distinguishing features, Bushell remained in Illinois. She made t-shirts for Andreessen and Bina to wear to their meeting with Clark (even though she was not invited) and kept in touch as Andreessen's relationship with the NCSA soured.[35] Legal disputes with the University of Illinois resulted in an early name change for the new company, leaving Mosaic behind to become Netscape Communications. Mac, Windows, and Unix browsers were written from scratch simultaneously with a focus on speed, stability, and new features.

Netscape began implementing its distinctive vision of a computer network, but it was still difficult to make the web accessible to those not technologically inclined. "To find one's way to the World Wide Web required setting up a [Serial Line Internet Protocol] or [Point-to-Point Protocol] connection, acquiring a modem and configuring ports, firing up the terminal window, typing arcane commands, using the file transfer protocol to download, install, and configure a web browser like Mosaic . . . And then, when finally on the web at long last, there was no easy way to discover what it was that others were even raving about."[36] Online services were not yet delivering their subscribers to the web. Prodigy was the first online service to provide web access, in 1995, and CompuServe and AOL followed soon after.[37] Netscape released Navigator 1.0 in December 1994, and it was an incredible success because it automatically loaded images directly into the text and had a point-and-click interface; it became known as a graphical browser. It also included the Secure Sockets Layer (SSL) protocol, which established an encrypted communication between browser and server that allowed (and continues to allow) for credit card numbers and banking information to be passed through the internet. Clark explained, "There needed to be an economic force driving [the internet], which meant that you had to be able to do business over the internet. So, we created a secure way of doing transactions. It's a bullet-proof rock solid way of authenticating two end points and having a secure connection. We invented that."[38]

It wasn't just that people—more and more of whom owned personal computers with less technical operating systems and simpler network connections—could view inline pictures and use a credit card. Rosanne Siino, Netscape's nineteenth employee, convinced the world that "the web is for everyone."[39] Under her watch, the company became a household name and symbolic of a new Silicon Valley. Siino became friends with Clark while doing public relations work for SGI in the late 1980s and later became the Vice President of Communications when he started Netscape. She had used Mosaic to access the web and found little to do or see, but Andreessen expressed a vision of what he thought the web could be. Siino sold that vision by selling Andreessen to those who had been interested in Clark.[40] Netscape received an extraordinary amount of attention from a fascinated and hungry population that increasingly owned home PCs.[41] Siino found the group of young guys inevitably compelling—and marketable. Andreessen became a Silicon Valley celebrity, posing shoeless atop a throne on the cover of *Time* in 1996, and Montulli was named *People* magazine's "Sexiest Internet Mogul" in 1999.[42]

Months after the initial browser release in summer 1995, the company went public—without profits but with anticipation and speculation that would come to define the dot-com bubble. The *New York Times* announced: "A 15-month-old company that has never made a dime of profit had one of the most stunning debuts in Wall Street history yesterday as investors rushed to pour their money into cyberspace."[43] Cailliau had hoped that the term "web browser" would by definition mean an editable application for navigating and accessing information on the internet. He argued, "The guy who brings out a passive browser spreads it faster, but it's not necessarily better for the user."[44] Cailliau had imagined an internet user who was much like Bush's Memex user: personal, reflective, and creative. Others, meanwhile, had a different vision of the web, as one full of consumers ready to embrace new forms of content, services, and experiences in easy, friendly packages.

Berners-Lee envisioned HTTP to be light and efficient, and the web of unidirectional hyperlinks to reach many. He designed the protocol to retrieve a document and disconnect in order for the web server to be free to respond to other users. In other words, the web is stateless by design, with each document or page request forming a discrete interaction. However, this retrieval system soon became understood as a "memory problem" for those working toward newly imagined potentials.[45]

In April 1995, Brian Behlendorf, lead engineer on HotWired.com and cofounder of Organic, Inc., wanted to talk about how users navigated websites.[46] He articulated the problem:

> There are a couple systems starting to be deployed now that attempt to gain information about "clickstreams." "Clickstreams" are the paths people take when they traverse your site—many content providers would find it useful to be able to detect common patterns or the effectiveness of various user interfaces. The problem is, of course, that HTTP is stateless, and beyond the hostname offers very little in the way of identification of unique "trips" through the content site.[47]

He sent the message to the www-talk mailing list, maintained by W3C, and wanted to discuss creating a new HTTP header called "session-ID." HotWired.com, part of *Wired* magazine, was the web's first online magazine. The site had been using access controls like HTTP basic authentication or the HTTP "From" header, but these requested an email address in response. Behlendorf included in his initial prompt that "any proposed solution *must* protect the anonymity of the user, for it's not really necessary to lose that when all that's cared about is unique sessions."[48]

Montulli replied that he had been doing "work along these lines" at Netscape. He detailed the set-cookie HTTP header and how it worked for the www-talk community.[49]

SYNTAX OF THE SET-COOKIE HTTP RESPONSE HEADER:

```
Set-Cookie: NAME=OPAQUE_STRING \
       [; expires=] \
       [; path=] \
       [; domain=] \
       [; secure]
```

SYNTAX OF THE COOKIE HTTP REQUEST HEADER:

```
Cookie: NAME=OPAQUE_STRING *[; NAME=OPAQUE_STRING]
```

The memory system chosen by Montulli worked as follows: when a client (e.g., user) requests an HTTP object (e.g., a web page), a server can send a piece of state information with the returned HTTP object (e.g., a cookie with a web page).[50] A cookie is a string of text that is placed into the memory of the browser, stored on a user's hard drive, and associated with a value available in a database available to the web server. Behlendorf responded to Montulli's cookie header, "A very mature proposal—now we need some test implementations :) Any volunteers? The browser should allow the user to NULL their session ID at any time (or turn off the functionality, even if by default it is on)."[51]

Although anonymous, the memory devised through cookie protocols is not interpretable by the user, but it can be controlled by the user. The term "cookies" was "very much in circulation in the original Unix lab," explained Doug McIlroy, who edited version 7 of the Unix manual published in 1979, where the term "magic cookies" first appears.[52] Bob Morris introduced and frequently referred to "cookies" as the "enigmatic bits of text or data that apparently came from nowhere, or that served as tokens that became meaningless out of context."[53] Morris used it as reminiscent of scattered crumbs.[54] The first cookie McIlroy remembers was "values of β may give rise to dom!" The Unix team justified funding at Bell Labs to typeset patent applications on a PDP-11, and one day that message was pronounced by the voice synthesizer ("values of beeta give rise to dom") on one of the consoles. The goofy phrase was incorporated as an error message in version 6. It appeared when trying to move files improperly. McIlroy recalls that, more generally, sometimes dummy files were created, and their presence or absence conveyed state information. Version 8 of the 1985 Unix programmer's manual included an entry that defined cookies as "a peculiar goodie, token, saying, or remembrance returned by or presented to a program."[55] The 1988 X Windows System manual states, "The magic cookie defined in a user's *.Xauthority* file is basically a secret code shared by the server and a particular user logged in on a particular display . . . Basically, under the magic cookie authorization scheme, a display becomes user-controlled."[56] PCMag.com's encyclopedia entry for "magic cookie" states, as of 2021: "A small data file passed from one program to another and sent back without change. Typically used in Unix systems, a magic cookie may be an identification token or password that activates a function. The 'magic' implies some obscure data known only to the software and not the user."[57]

Web cookies are an iteration of magic cookies and operate similarly, with the same characteristics. Future HTTP requests made by the client to the server will include the most recent value of the cookie pulled from the client's browser storage and sent back to the server. Rather than "magic," it is referred to as "opaque" memory, meaning it is "opaque to the user agent and may be anything the origin server chooses to send . . . 'Opaque' implies that the content is of interest and relevance only to the origin server."[58] Despite dropping the title, HTTP cookies are still magic.

Although Behlendorf initially stated that users would need to both stay anonymous and exercise control over cookies, few mentions of the latter were made in the remainder of the discussion on the www-talk mailing list. Identification itself was considered a privacy issue only in "extreme cases," but control was a means of providing privacy, educating and empowering new users/consumers, and recognizing market forces.[59] Transparency is an important component to control, however. In his message arguing to keep the discussion alive, Marc Hedlund, working as Director of Engineering at Organic, wrote, "A while back someone asked what was wrong with the Netscape Cookie proposal. A long delayed response: it doesn't tell the user what it's doing."[60]

Koen Holtman, one of the few Europeans engaged on the list, but one of the many on the listserv looking beyond the specific issue that Behlendorf brought up, cared little about improving site statistics for marketing. Holtman wanted to go beyond a user filling in a simple web form and clicking submit and hoped to create a two-way interactive dialog between the user and web server. He suggested a browser interface that might provide user control; it was inspired by the selections users made to download images, which on early networks could take a significant amount of time:[61]

+——————————————————————————————+

HANDLING OF A) 'STATEFUL DIALOG' SESSION-ID REQUESTS:
   ( ) Always honor request
   ( ) Always honor request if it was done in a response to a form submission (POST).
   (*) Ask once for every site, use reply in later sessions
   ( ) Never honor request

GENERATE B) STATISTICS-ENHANCING SESSION-IDS:

( ) Yes

(*) No

+―――――――――――――――――――――――――――――+

where the (*) are the default settings

In the same message, Holtman asked, "What happens if the makers of commercial browsers get interested in expanding their business to making web statistics packages, and start shipping browsers with default settings . . . b) statistics-enhancing session-ids: (*) Yes ( ) No?," which is, of course, exactly what happened.[62]

The debates included further reminders of the reality of "the market" as an inevitable force pushing for both metrics and new functionality. These reminders did not always come from commercial actors. Robert Robbins, who was then a faculty member directing informatics work on the Genome Database for the international human genome project at Johns Hopkins, had set up a section of his personal web page to share papers with colleagues around the world. Robbins attempted to draw an analogy between the session-ID conversation and caller ID on telephones, arguing that the problem of the competing desires spurred by the development of caller ID in the late 1980s—wanting to identify callers, wanting to not be identified as a caller—had been met by developing protocols that allow callers to block sending their information and recipients to block calls without caller ID; after these protocols were established, it was up to the market to sort out the rest. Robbins warned:

> Technical solutions that are based on social judgements have problems, the biggest of which is their lack of generality. Computer systems that seek large market share should not attempt to behave in a paternalistic manner, deciding for users what their legitimate needs are and meeting those needs (and only those needs) . . . Systems that are paternalistic . . . are not likely to gain large market share and may indeed lose market share over time to systems that allow the user to do what he wants without requiring him to do it only that way. Market pressures demand session-IDs, both for maintaining states and for user tracking. Either session-ID capability will be implemented sensibly or it won't, but in no case will its emergence be prevented . . .[63]

The "market" would demand tracking and its form could be worked out by users.

In his first message, Behlendorf explained that he had people asking for clickstream data and statistics, and a theme of marketing demands came in and out of the threads that followed. Wrapping up his final call for consensus or an end to the discussion, Behlendorf wrote, "Anyways, at this point I'm ready to throw in the towel and just go back and tell the people screaming for this kind of functionality and go 'look, without intrusive measures like password protection or broken mechanisms like session-ID-URL-munging or heuristics which work one day and not the next, I just can't get that info for you with any reasonable accuracy.' And then we'll see which path they choose."[64] Largely driven by Netscape's implementation, consensus formed later around cookies, not site log-ons or session-ID URLs, as a means to give consumers what they allegedly wanted.

ADVERTISING AS INFRASTRUCTURE

Even with the popularity of Netscape Navigator, it was not clear how Netscape would make money. A range of revenue sources were being pursued by internet entrepreneurs in 1994, including utility charges for connectivity; access to the web and email; subscriptions and premium fees for specialized information and entertainment; licensing schemes for browsers, search tools, and software libraries; and advertising. Netscape eventually brought revenue from enterprise software sales that included server packages; browser, email, and custom software; annual support contracts; and ad revenue derived from traffic on the portal and distributor bundle packages on operating systems. But it wasn't clear how or to what extent web companies could be profitable or what industries they would support, need, or threaten.

Kevin O'Connor and Dwight Merriman considered money the primary problem the web needed solved.[65] How would people make money on the web? As they brainstormed a new venture in Atlanta, the two asked what would serve as the "plumbing" and "electricity" of the web.[66] They continued to come back to the inevitability of advertising and started the Internet Advertising Federation in 1995. They set out to build a kind of economic infrastructure for the decentralized network of the future that would take on the walled gardens of online service providers, namely AOL, CompuServe, and Prodigy.

Both men were trained engineers, O'Connor with a degree in electrical engineering and Merriman with a degree in computer science. But

O'Connor took up learning the business of advertising and headed to the library to check out famed "master strategist of direct marketing" Edward Nash's books.[67] O'Connor became convinced that the web could provide the measurements, tracking, and targeting at the core of advertising, and Merriman worked to deliver on the technical side. As O'Connor studied contemporary advertising, reading marketing news at the library and looking for potential competition, he came across an announcement for DoubleClick out of Poppe Tyson. It was a new division in the traditional advertising firm Bozell, Jacobs, Kenyon & Eckhardt designed to sell advertisements on clients' websites. It turned out the advertising firm needed Merriman's technology, and the two companies combined forces under the name DoubleClick. O'Connor and Merriman moved to New York to plant a flag in what was becoming known as Silicon Alley.

Advertising as infrastructure was not a new idea in media or communication, and its powerful legacy as essential to free or subsidized information services from newspapers to television contributed to a consistent and convincing argument for a "free" internet. If advertising kept the lights on and quickly delivered new technologies, this thinking went, then innovation policies should foster, not hinder, building the advertising infrastructure. Although the engineers saw new potential for advertising on the web, the vital job of selling the internet and online advertising to traditional advertising players who were more comfortable with the metrics of radio, television, and print fell to Wenda Harris Millard. She was a founding member of DoubleClick's executive team.[68] There's a saying in advertising: "If you don't know Wenda Harris Millard, you're not in marketing."[69] After twenty years in traditional media, Millard knew everyone and quickly established the company's legitimacy on Madison Avenue, translating online services into terms the industry used and educating traditional groups. Double-Click became the dominant player in online advertising; a small group of like-minded companies were founded at the same time, but none would dethrone it.[70]

These other internet advertising companies came to see the potential of an ad-supported web from different perspectives. Gil Beyda and Dave Morgan created Real Media at the beginning of 1996.[71] Beyda was a California developer who started his first tech company, Mind Games, in high school while working at a local computer store called Computers Are Fun. Morgan had worked as an attorney for a newspaper trade association with a commercial arm selling ads to its members.[72] Engage was another early online advertising company, cofounded by Daniel Jaye, who trained in astrophysics and worked on computing for scientific projects at Bell Labs and later managed databases across Fidelity Management and Research. He created Engage in 1995 as part of College Marketing Group Information Services' (later CMGI) push to cultivate the internet.[73] DoubleClick, Real Media, and Engage all sought to provide advertising to support independent sites on the new decentralized web through a centralized network of advertisers. The companies drew distinctions between themselves, to some extent: namely, RealMedia was initially designed to provide ads to small news sites, developing relationships with publishers, whereas Engage focused on a direct marketing strategy, advertising directly to targeted individuals. All have claims on originating targeted behavioral advertising, but when the dot-com crash ripped through the industry, what emerged was a handful of consolidated companies that would "target users, not pages." Users would be tracked across sites that would sell access to their "unused inventory"—that is, their user data.[74]

Using computers to better target direct advertising had raised concerns before, however. The "consumer-database" industry of the 1980s and 1990s confronted the privacy issues of the previous decade and of the decades come. A *Wall Street Journal* article from 1991 asked:

> Want to send junk mail to short fat guys with glasses? The consumer-database industry has the service for you. In their eternal search for intimate data about Americans, a few firms are buying state driver's license records to get such data as consumers' height, weight and use of "corrective lenses." Even in the omnivorous mailing-list industry, the use of height and weight data is controversial. Most drivers who register for a license have no idea their personal information finds its way into the hands of marketers. Two of the biggest consumer-database companies . . . don't sell height and weight data because it's too personal.[75]

In the late 1980s, the public and private sectors promoted "computer matching" as consumer-friendly and cost-reducing, but still triggered privacy concerns.[76] When Lotus Development Corporation revealed its newest product, the Lotus MarketPlace, at the MacWorld Computer Expo in 1990, it targeted small businesses with the aim of increasing sales through direct marketing mailing lists that could be run on a personal desktop computer. The goal was never realized, because Lotus shelved the product in response to public outcry over privacy concerns.[77]

Still, the web provoked new possibilities for advertising. A 1996 *Wired* article promoted a new "webonomics," stating, "Not someday, today—advertising on the Web makes economic sense. You just have to forget everything you ever learned about the business."[78] Advertising could only serve as infrastructure for the commercial internet if websites and services could get a chunk of advertising budgets. Online advertising companies argued they had better data and could more precisely target potential customers than other forms of marketing. The internet needed the advertisers, and the advertisers needed the data—or so the argument went. Advertising on predecessor networks like Usenet was controversial.[79] Behlendorf explained that the difference was the nature of the networks. The Usenet network was "mutualized" and operated as a commons, whereas the web was delivered to users by a commercial service provider, presented through the package of a corporate home page, and conceived by those working on it as a place for media consumption, just like any publishing enterprise that placed advertising next to content.[80]

Despite this difference in network composition, many on the early web had no interest in its commercial aspects. Artists like Mark Amerika played with state to create novel forms of visual art and poetry. One of his most groundbreaking pieces is GRAMMATRON, which he developed as a creative writing graduate student at Brown. While hanging out in van Dam's Graphic Visualization Lab (Ted Nelson's haunt in the 1960s), Amerika met a student named danah boyd, who wrote the first set of JavaScript cookies that would alter the reader's path through the GRAMMATRON narrative.[81] The piece was exhibited in museums and covered by the *New York Times* and *Wired*.[82] It remains freely available online and does not require upkeep because it relies on cookies. The early web was not all metric-driven or algorithmically delineated. It was carefully curated, intimate, and quirky—it was cool. Coolness drove a usable, social, and commercial web.[83] The Cool Site of the Day, referred to as CSotD and called the "arbiter of taste," made its creator, a Virginia ISP operator named Glenn Davis, an internet celebrity. CSotD was made up of "links to movie trivia and webcams trained on fish tanks, mortgage calculators and email syntax guides, clip art collections, and NASA images of outer space. You could play a hand of blackjack, peruse Bartlett's database of quotations, pose a question to the Magic 8 ball, and answer an online personal ad. There were satellite weather maps and feminist music communities, sci-fi enthusiasts and BDSM aficionados, geek subcultures and mainstream marketers—a cacophony of weird things, helpful tools, unexpected resources, and diverse voices made to seem suddenly close, personal, and rendered in rich detail."[84] Nonetheless, the Cool Site of the Day had advertising—a Cool Sponsor of the Day.[85]

Reasserting this need to sort out economic stability, Martin Nisenholtz, who founded the Interactive Marketing Group (IMG) at the advertising agency Ogilvy & Mather Worldwide, wrote for *Advertising Age* (covered by the *New York Times*) in 1994:

> Radio, telephone, and TV networks were all born primarily as commercial services . . . If you were to advance the theory (in, say, one of the alt.newsgroups) that the purpose of the Internet was to make money, you would probably get flamed . . . No one owns the Internet. There is no single governing body that controls it . . . Everyone would probably like to see the Internet become a financially self-sufficient entity. The question is how.[86]

Nisenholtz worked on teletext as a form of interactive consumer media in analog broadcast television spectrum (and later cable), as well as with videotex to bridge the telephone with the television using a decoder box.[87] Calling for guardrails around a set of guidelines for the web, Nisenholtz proposed six: 1) intrusive email is not welcome, 2) internet consumer data is not for resale without the express permission of the user, 3) advertising is allowed only in designated newsgroups and list servers, 4) promotions and direct selling are allowed but only under full disclosure (consistent with the rules developed by the Direct Marketing Association for "analog merchants"), 5) consumer research is allowed only with the consumer's "full consent" (which meant "ready and easy access to information outlining the uses and implications" of participation), and 6) internet communications software must never hide concealed functions. In what may represent a first version of a web privacy policy, these tenets speak to a moment of internet culture when the internet "consumer" was being carved out by internet marketers.[88] Could the advertising infrastructure of the web be built with these kinds of restrictions on data?

American online advertising companies quickly expanded to Europe (as well as Australia and Asia), bringing with them the idea of an ad-supported free web. They faced few domestic competitors abroad, largely because the national markets in Europe were themselves too small, with too few national websites, to support themselves.[89] "To get there first is an important part of our strategy," said DoubleClick's O'Connor in 1998, "We're forced to become a global company."[90] Recall that online service providers and their advertising and/or subscription models varied from videotex services with mixed public–private arrangements like the French Teletel (as discussed in the previous chapter, a system based on an X.25 network, and the famous French Minitel) and the German Bildschirmtext (later supplanted by the Deutsche Telekom's T-Online internet service), to AOL Europe and the United Kingdom's Freeserve. That said, advertising had been an integral part of globalization in the 1980s and had become a massive byzantine business. Because Europe was a relatively fragmented market, the strategy for the American companies was to find a local marketing partner who knew the domestic advertising landscape.[91]

In short order, European online advertising became intertwined with American companies through mergers, deals, and buy-outs. Countries with newly privatized telecommunication systems took up the economic logic of advertising web infrastructure. For example, Michael Stusch and Michael Schultheiß founded AdLINK in Germany in 1998, 80 percent of which was acquired by the American CMGI in 2000. AdLINK used technology from AdForce, an American company acquired by CMGI in 1999, to deliver ads until the company was shuttered in 2000. In 2001, when it was "rightsizing," DoubleClick sold its European media sales business to the German AdLINK, which continued to use DoubleClick's products and services. In 2003, United Internet Media, part of the German internet service provider United Internet, joined with five other European media networks to create AD Europe. And in 2004, DoubleClick sold its shares in AdLINK to United Internet, the ad network's majority shareholder. In 2009, Hi-media out of Paris bought the display unit of AdLINK. Hi-media was started in 1996 by twenty-four-year-old Cyril Zimmermann and became a Pan-European advertising network in the early 2000s, when Engage, RealMedia, and DoubleClick had already entered the market. As web advertising infiltrated, educated, and persuaded the traditional, "highly complex global service industry" that had become advertising by the 1980s, it was largely an American force that proliferated through the global advertising apparatus.[92]

Pushing for an ad-funded commercial web was perhaps a subtle version of Americanization through a strategy of "unilateral globalism."[93] The Clinton administration was less subtle. The Clinton administration wanted a global internet—an internet funded by the US government and maintained by American companies. The administration argued that no country should be in control of the internet, though; business should take the lead and consumers would express their demands. Then-presidential and vice-presidential candidates Bill Clinton and Al Gore (a techy himself) had campaigned on Atari Democrat themes and courted Silicon Valley with the help of established tech company leaders, who contributed to the campaign's broadband technology policy.[94] Once in office, the Clinton administration promised significant government support of and investment in Silicon Valley's computing technologies, and the National Information Infrastructure (NII) plan was written by appointees in the DOC, not in the Federal Communications Commission (FCC).[95] Enlisting the FCC would have signaled a different set of priorities like access and media responsibility and would have been set forth by expert communication lawyers, economists, and engineers working in a richly regulated and fee-based sector. Shifting NII activities into a new realm of expertise and regulatory culture signaled a new phase for the web, and the plan made it clear that the internet was central to US economic policy.[96] There was little resistance to such a shift. Tech issues were not a priority for most Americans, and when the NII plan was released in September 1993, the public didn't take notice.[97]

In 1997, Clinton directed the Secretary of Commerce to privatize the domain name system, increasing competition and international participation.[98] The DOC released its "Proposal to Improve the Technical Management of Internet Names and Addresses," which was a plan for international internet governance.[99] It had four principles: 1) the private sector should lead development of and set standards for electronic commerce, 2) governments should avoid disincentivizing or inhibiting electronic commerce with regulations, 3) governments should instead encourage self-regulation, and 4) governments should cooperate and harmonize among themselves to foster electronic commerce.[100] The proposal uses a mix of references to "users" and "consumers," sometimes interchangeably: "The Internet succeeds in great measure because it is a decentralized system that encourages innovation and maximizes individual freedom. Where possible, market mechanisms that support competition and consumer choice should drive the technical management of the Internet because they will promote innovation, preserve diversity, and enhance user choice and satisfaction."[101]

Foreign governments and intergovernmental bodies complained about international oversight.[102] The European Commission responded, "[F]uture management of the internet should reflect the fact that it is already a global communications medium and the subject of valid international interest."[103] Behind the scenes, the Clinton administration had been told, "the Europeans are mad as hell."[104] France pushed back the hardest, wanting to deemphasize e-commerce and talk about content issues and human rights. Europeans were not convinced about the United States' self-regulatory approach and thought it left too much power in US hands, but the Clinton administration argued that anything less was isolationist and that domestic regulation would harm citizens around the globe.[105] This moment, orchestrated by the Clinton administration, transformed users into consumers and, over the course of the late 1990s, it turned internet privacy into consumer privacy.[106]

CONSUMER PRIVACY

The consumer is a relatively new figure. According to historian Peter Stearns, "Consumerism describes a society in which many people formulate their goals in life partly through acquiring goods that they clearly do not need for subsistence or for traditional display. They become enmeshed in the process of acquisition—shopping—and take some of their identity from a procession of new items that they buy and exhibit. In this society, a host of institutions both encourage and serve consumerism."[107] Although hints of consumerism have existed throughout time, modern consumerism took shape in Western Europe three hundred years ago.

Consumer societies can be found in 1700s Britain, France, Germany, the Low Countries, and Italy, in a distinct moment when shops proliferated and marketing methods expanded dramatically. The expansion of consumerism involves the transformation of goods like soap and coffee from luxury to necessity. European consumerism spilled over into colonial America, which experienced some commercial wealth from demand for tobacco, cotton, and sugar. However, Stearns describes the US "only as an imitator, even though the nation would ultimately develop a world lead in some facets of consumer standards."[108] American consumerism eventually followed European fashion, food, and comforts. But this took many years, as most Americans were initially traditional rural barterers. The US caught up quickly in the 1800s, however, embracing French fashion and food and British shops and habits. By the 1880s, the student had become the master, due in part to the huge and growing single national market of the United

States. Consumerism took off in both regions, with varying stops and starts for events like wars and natural disasters.

After World War II, what historian Lizabeth Cohen calls the "consumers' republic" took shape: a United States in which the consumer is the quintessential citizen and free enterprise is the ideal democratic model.[109] Cohen describes how during the New Deal and World War II, "consumer citizens" saw their rights as essential to promoting the common good; after the war, however, the "purchaser consumer" came to dominate, and understood their only obligation as citizens to be increased consumption and, in turn, American prosperity.[110] During this postwar period, the US State Department collaborated with the Advertising Council to "sell Europe on free enterprise," as propagated through the Marshall Plan.[111] The Advertising Council was "staunchly and unambiguously in favour of free market capitalism as the only route to democracy, and it used its extensive resources and privileged relationship to the White House to spread that message."[112] The "Overseas Information" campaign, originally called "Selling America Overseas" in 1948, did not enjoy the same enthusiasm that selling America to Americans received.[113] Although companies like Westinghouse creatively incorporated the suggestions made in *Advertising: A New Weapon in the World-Wide Fight for Freedom; A Guide for American Business Firms Advertising in Foreign Countries* to support the "newest weapon of democracy," the effort fizzled out in 1949.[114] The campaign to sell America to Europe was considered a failure due to a general lack of buy-in from advertisers, possibly because of Europe's small and underdeveloped advertising sector, paper shortages, and language barriers at the time.[115] The Marshall Plan also met with great disparity in messaging, reception, and resistance to the consumers' republic: West Germans reoriented around consumer democracy to rebuild; the English Labour Party communicated the Marshall Plan not in terms of mass consumption but through a message of fair means and austerity; and the French, armed with a formidable arsenal of critiques of consumer capitalism, revolted against attempts to embrace consumerism in exchange for their dependence on their unions.[116] When the European welfare states came under attack in the 1980s, many expected dramatic Americanization to follow, but the more expansive social citizen of Europe was increasingly attached to European social charters.[117]

Although historical debates over consumerism in domestic and international policy eras are ongoing, most accounts find marked increases in consumerism after World War II and pinpoint the formalization of the consumer as a political subject as occurring in the 1960s.[118] President Kennedy delivered his "Consumer Bill of Rights" speech to Congress in 1962:

> Consumers, by definition, include us all . . . The Federal Government—by nature the highest spokesman for all the people—has a special obligation to be alert to the consumer's needs and to advance the consumer's interests . . . If consumers are offered inferior products, if prices are exorbitant, if drugs are unsafe or worthless, if the consumer is unable to choose on an informed basis, then his dollar is wasted, his health and safety may be threatened, and the national interest suffers. On the other hand, increased efforts to make the best possible use of their incomes can contribute more to the well-being of most families than equivalent efforts to raise their incomes.[119]

Kennedy's consumer was not limited by monopoly powers, but rather plagued by choice: "The typical supermarket before World War II stocked about 1,500 separate food items—an impressive figure by any standard. But today it carries over 6,000. The housewife is called upon to be an amateur electrician, mechanic, chemist, toxicologist, dietitian, and mathematician—but she is rarely furnished the information she needs to perform these tasks proficiently."[120] He outlined four rights for this consumer: the right to safety, to be informed, to choose, and to be heard.

Kennedy's "Consumer Bill of Rights" resonated internationally. In 1975, the European Economic Community launched its preliminary program for consumer protection and information policy, which included five fundamental rights.[121] Recall that the EEC was created in 1957, when France, West Germany, Italy, Belgium, Luxembourg, and the Netherlands signed the Treaty of Rome (renamed the Treaty on the Functioning of the European Union, TFEU) to establish a harmonized single market. The 1975 program declared, "The consumer is no longer seen merely as a purchaser and user of goods and services for personal, family or group purposes but also as a person concerned with the various facets of society which may affect him either directly or indirectly as a consumer."[122] This consumer had a right to health and safety, to the protection of economic interests, to compensation, to information and education, and to representation. The right to economic interests included prohibitions on false advertising: "An advertiser in any medium should be able to justify, by appropriate means, the validity of any claims he makes."[123] Article 169 of the TFEU enables the

EU to protect consumers. In the 2000 Charter of Fundamental Rights of the European Union, a "high level of consumer protection" became a fundamental right.[124] Consumer protection efforts happened in parallel with the data protection and privacy initiatives described in the chapters preceding and following this one. European consumer privacy is like American hate speech—it doesn't legally exist.

In the US, the FTC, which has become a kind of de facto federal personal data regulator, was central to creating the Privacy Consumer out of consumer privacy. In his detailed history of the FTC, legal scholar Christopher Hoofnagle defines privacy "broadly as the FTC's consumer protection activities relating to regulation of data about people" and explains that the FTC is a complex, agile, innovative agency—even though it is frequently criticized for not doing or being enough.[125] How did an agency rooted in public concern over monopolies and trusts, with responsibility "for over seventy laws, concerning fraud in college scholarships, false labeling of dolphin-free tuna, health warnings over cigarettes, the labeling of furs and wool products, and even the sanctioning of boxing matches," come to be in charge of personal data in America?[126] Advertising.

Hoofnagle explains, "Instead of becoming an inflexible institution, [the FTC] changed as marketing changed. As marketing evolved from print to radio to television to the internet, the FTC retooled its investigative practices."[127] The agency expanded beyond its initial reach as established by the Federal Trade Commission Act of 1914 (legislation that passed on a wave of anti-business sentiment and focused on anticompetitive behavior) by actively pursuing deceptive trade practices in the 1920s and 1930s.[128] The Wheeler–Lea Act of 1938 was part of the second phase of consumer history; it passed in an almost unimaginable context, when the only regulation of products was administered through labeling.[129] In their 1933 book *100,000,000 Guinea Pigs*, Arthur Kallet and F. J. Schlink spurred legislative action as they reinforced disdain for advertising.[130] They recounted story after horrifying story, like that of Kopp's baby soother, which killed nine toddlers; penalized with a $25 fine, the company simply disclosed that its product contained morphine.[131] The advertising industry was new and, on the heels of a 50 percent decline in ad spending during the Great Depression, it was resistant to being hobbled by regulation.[132] Although the Wheeler–Lea Act provided new and important powers to address false advertising, the ad industry breathed a sigh of relief in the years that followed, and indeed a Gallup poll in 1940 found that 59 percent of people wanted more advertising regulation.[133]

The history of US consumer privacy revolves around regulating consumer information that is used for direct marketing to consumers and to extend credit to consumers. These two purposes were frequently discussed together. The first FTC action to protect consumers from unfair information collection involved a company that helped creditors find debtors using mail in 1951. Gen-O-Pak sent debtors postcards that said a package was waiting for them and would be sent if they replied with the necessary information.[134] Arthur Miller, the legal scholar who played an important role in articulating the data subject in chapter 2, wrote that credit bureaus had been using computers since 1965 and warned that "given the massive investment required to computerize a large credit data base and the technology's ability to manipulate bits of information in unique ways, the temptation to use the data for non-credit-granting purposes will not be easy to resist . . . such as generating a special mailing list containing the names of consumers with certain characteristics who might be interested in a particular product."[135] Miller knew credit data cost something to collect, clean, and process and that it had value beyond credit determinations. During House of Representatives hearings on commercial credit bureaus in 1968, law and policy scholar Alan Westin described how credit reports were disclosed to non-credit granting entities, including the FBI and labor unions. He even got his hands on a report—just by asking for it.[136] Representative Gallagher (the same congressman whose fears of the "computerized man" were described in chapter 2) followed up. He asked, "Do you see any possibility of limiting credit information to the sole use of determining credit, or must we resign ourselves to another of the seemingly endless cases of data gathered for one function and used for a totally different purpose?"[137] Gallagher expressed concern to Westin over children and adults who participated in "the dating game business through computers" and were subsequently put on a list to be mailed "pornographic literature." Westin responded, "I think at some point a market mechanism must be established. I would like to see people informed whenever the presence of their name on one list gets them on another list. Again, the individual should have the right to decide. I think it is so technologically simple for things like this to be done that we haven't even begun to explore the possibilities."[138] Westin envisioned a neat technological solution that would inform people and launch this market: the computer would be programmed to print a postcard that would then be mailed to the consumer.

Although extending credit went alongside mail marketing in the conversation about regulating the collection of consumer information, it received different legal treatment. Administered by the FTC, the Fair Credit Reporting Act (FCRA) passed in 1970. In his history of consumer surveillance told from the perspective of the credit reporting industry, Josh Lauer traces investigations into consumers starting in the early 1800s. He explains: "By the time Americans awoke to the problem of consumer privacy . . . it was too late. A national credit reporting infrastructure was firmly in place and in the early stages of computerization."[139] In the late 1960s, hearings revealed that credit information was being shared with many entities for many reasons, but as long as records could be kept separate, the main problem that policymakers acknowledged was dealing with errors.[140] Although credit bureaus presented new industry guidelines in 1969 that would allow consumers access to their files and the opportunity to correct mistakes, as well as to delete adverse information after seven years, complaints and errors continued, and were shrugged off as anomalous and minor inconveniences.[141] Unconvinced by industry efforts, lawmakers passed the FCRA to limit the procurement of credit reports for the problematically broad "legitimate business need" and required that credit bureaus disclose only "the nature and content of all information," not the actual file. At the time, there were over 2,000 credit bureaus in the country, many of which were run by local credit managers who had no intention of automating.[142] Because of this state of the industry, credit data sharing was not a major issue, unlike the broader concerns about data sharing that dominated the *HEW Report*. But by the end of the 1980s, there were only three credit bureaus, and they were expanding into database marketing for direct mail. After tangles with the FTC over this new venture, the bureaus kept data for credit reporting (i.e., payment histories) separate from consumer identification information for marketing. The data used for direct mail was subject to lighter legal treatment, despite Gallagher's efforts.

OPT-OUT

Direct mailers using databases of consumer information wanted to avoid the type of regulation that credit bureaus got. The legacy of opt-out began in 1971 with the Direct Mail Advertising Association's (DMAA) flagship program: the Mail Preference Service.[143] This was the same year that, after 180 years of congressional ratemaking, the US Post Office became a government corporation called the US Postal Service, in what was "arguably the most fundamental restructuring of a major federal agency in American history."[144] The DMAA established the Mail Preference Service after bills restricting direct mail lists were introduced in 1969 and 1970.[145] In a hearing on these bills, New Jersey Representative Gallagher stated:

> Nothing more clearly illustrates this [danger to privacy and the Bill of Rights] than a recent event in Chicago which is directly relevant to these hearings. A suit has been filed by Encyclopedia Britannica against three former employees charging that they copied computer tapes. These tapes contained the names and addresses of two million Encyclopedia Britannica customers and were sold to mailing list brokers. One broker, who had received the stolen property, then sold a list containing 800,000 of those names to Curtis Books, Inc., a competitor of Encyclopedia Britannica. I believe this shocking example typifies the necessity for the Congress of the United States to begin to understand how vulnerable to abuse the computer is, and how its willing tool, the junk mail industry, welcomes and actually encourages such abuse. At least, it has no scruples about buying any collection of names, and the humans to whom those names refer have utterly no control.[146]

Gallagher told another story about how a woman's pregnancy was revealed to her family when diaper coupons were mailed to her home after a lab sold her information.[147] He insisted direct mailers know far too much about Americans, while little was known about the industry. The hearing also included a *New Yorker* article explaining how list brokers sold information to target addresses from census data: "As it turns out, one of the minor disadvantages of being a Negro slum dweller in the United States is being deprived of a fair share of orange-juice coupons."[148] Data breaches, sexual privacy, informational asymmetries, and racial bias were among the issues with direct marketing that members of Congress highlighted in 1970.

In response, representatives of the DMMA argued that direct mail was an important form of advertising, that it had no real harm, and that the association was in favor of limiting the distribution of pornographic mail. But it did not mention mechanisms of control until 1971 when, as bills to ban pornographic mail floated around Congress, the DMMA presented its new program to the House Appropriations Committee:

> We recognize that in this uptight world there's a percentage of people who don't like anything. Since a key characteristic of our medium is selectivity, DMAA is pioneering a forward-thinking program to help consumers. We call it the Mail Preference Service (MPS), whereby we ask our cooperating members to remove from their mailing lists the names of those who seriously object to all advertising. Curious point: despite the alleged hue and cry that this mail is unwanted—when given the opportunity pilot tests confirm what has been indicated all along: most people prefer to receive advertising mail. They want to make the individual choice each time whether they are interested in a particular sales offer.[149]

The organization described how it test launched and publicized the MPS in newspapers but received very few inquiries about the program. With fewer than fifty forms returned with a preference to be placed on the "off" list, the organization concluded that people want to receive advertising in the mail.

The 1973 *HEW Report* directly addressed mailing lists, giving them their own appendix, yet it likewise deemed them benign.[150] Although the growth of direct marketing is tied to computing and especially printer technologies, at the time, computer equipment was large, cumbersome, and labor intensive, so maintaining detailed addresses and data for individuals was not worth the effort or resources except for very large direct mail marketers (mostly successful national magazines).[151] The HEW Committee found no evidence that direct mail advertising was "anything more than an annoyance to a small part of the population" and that their principles (recall from chapter 2, the basic fair information practices provided that data processing should be known, transparent, purpose specific, correctable, and reliable) did not seem applicable.[152] Noting that "an underlying function of the Advisory Committee's recommended safeguards is to provide effective feedback mechanisms that will help to make automated personal data systems more responsive to the interests of individuals," the committee agreed that direct marketers already "concentrate almost obsessively on methods for maximizing response and minimizing complaints."[153] The HEW Committee argued that the DMAA needed to publicize the MPS more prominently. They also suggested that, if mailing lists become a problem, a check box could be provided on forms like applications or purchases that would allow the recipient to grant consent for secondary use, in line with *HEW Report* principles.

When the Privacy Protection Study Commission (PPSC, recall from chapter 2) began holding hearings in 1975, it sought to "answer the question raised by the [Privacy Act] . . . 'whether a person engaged in interstate commerce who maintains a mailing list should be required to remove an individual's name and address from such a list upon the request of that individual.'"[154] Because of public interest, the PPSC dedicated three days of hearings to rights related to mailing lists. More so than the HEW Committee, the PPSC discussed the postal service's economics and its reliance on direct mail. When justifying its recommendation to not legally enforce an individual's request to be removed from a mailing list, the Commission explained the historical importance of mail to the freedom of expression, charity, and elections, and the postal service's reliance on direct marketing. It further explained that direct mail was important to the national economy and that putting a removal system in place would increase the costs of operating the mail and drive advertisers to the telephone.[155]

In the end, privacy in terms of mailing lists was considered sufficiently protected by the "negative check off systems" implemented by individual entities like American Express and Computerworld, which managed their own mailing lists, and the organized "delisting" put in place by the DMAA.[156] Organizations that made personal information available to third parties were to provide detailed notice and implement removal processes with their partners. Such organizations were not to make personal information available to third parties beyond the context of the original purpose, unless they informed "customers" and gave them an opportunity to indicate that they did not want their names used for such secondary purposes. The PPSC did not recommend any legal measures, and instead recognized the "difficulty and the undesirability of forcing record keepers to assume that responsibility" and explained that "because so many appear to be willing to assume it voluntarily, the Commission believes that voluntary implementation is likely to be a successful as well as adequate solution to the problem."[157]

Although direct marketers avoided privacy regulation by adopting this opt-out system, they didn't avoid FTC enforcement entirely. In 1971, the FTC ordered Metromedia to cease its deceptive practices after the company mailed out surveys that rewarded respondents with a chance to win "1,111 magnificent gifts."[158] The three-page questionnaire was accompanied by a letter stating, "There is nothing to buy and we should assure you that no salesman will call on you." In fact, the survey was used to compile mailing lists to generate product sales.[159] Before the internet, the FTC interpreted norms of confidentiality in relation to selling consumer information as a strong privacy rule and focused on opt-in consent.[160] But the Reagan administration held the FTC up as an example of a "nanny state" run wild, rallying several business interests and critics around the proposed regulation of children's advertising.[161] The Reagan era at the FTC was contentious and intended to respond to what the administration considered the agency's excessive activities in the 1970s, but also shifted its paradigm away from lawyers focused on bad marketplace actors and to economists focused on the consumer.[162] President George H. W. Bush appointed Janet Dempsey Steiger, neither a lawyer nor economist, as the first chairwoman of the FTC in 1989. Steiger was so successful that President Clinton broke with previous political patterns and kept her in the position when elected. Clinton later appointed the agency's most experienced chair, Robert Pitofsky, in 1995. Pitofsky set up the FTC's privacy program. Hoofnagle describes how the FTC moved into online privacy under their leadership:

> [The FTC] brought its first internet case in 1994, before most Americans were online, and when many thought that the internet was just a fad . . . Early privacy matters focused on enforcing privacy representations, but companies learned quickly to not make specific promises or to write vague privacy policies. As companies changed their strategy, the FTC shifted to an approach of enforcing consumer expectations. The FTC borrowed heavily from its expertise in recognizing and remedying false advertising. Just as it did a generation before when evaluating mass-media advertising, the Commission looked to website design, settings, and even informal remarks by employees as informing consumers' perceptions of privacy promises.[163]

Based on the FTC's assessment, by the 2000s websites themselves were ads selling their content, services, and communities, and all Americans online were consumers of privacy. To protect consumers online, the FTC focused on eliminating unfair practices through a case-by-case enforcement strategy to protect consumers online. It has never attempted a broad administrative rulemaking for online consumer privacy, outside of children's privacy and areas like telemarketing.

These other areas provide insightful distinctions between opt-out systems. The Telephone Consumer Protection Act (TCPA) of 1991 prohibited sending unsolicited advertisement faxes (junk faxing) and restricted telemarketing hours and practices. The FCC was charged with making specific rules for the TCPA and required each company to maintain its own do-not-call database. This didn't work. People were still receiving unwanted phone calls and, in 2003, the FTC—not the FCC—was tasked with establishing the National Do Not Call Registry. Consumers could opt out of receiving marketing calls by adding their name to just one list maintained and enforced by the FTC. For marketing via phone call or fax, policymakers were more concerned with running afoul of the First Amendment and less concerned with maintaining the economic infrastructure of the network.

More advertisers became aware of the web through mainstreaming efforts by companies like DoubleClick and Netscape, but the public became aware of cookies from a British *Financial Times* article published in February 1996 that focused on transparency and consumer choice. After introducing cookies, the story explained, "Most extraordinary of all, this information can be stored on customers' own PCs without their knowledge. It can be kept in a form so that only the company that collected the information can benefit from it . . . Moreover there appears to be only one way to disable the facility: by manually amending or deleting the COOKIE.TXT file containing all the cookies."[164] It further, though incorrectly, explained, "They do not allow one company to snoop on another, and they gather only information about consumers' behaviour at a single company's Web site, or information that customers themselves volunteer."[165] Soon, the technical debate around the advertising infrastructure and the role of consumers in supporting that infrastructure became a public debate about opt-in versus opt-out.

David Whalen's FAQ for Cookie Central reveals this stark shift. Cookie-Central.com, an odd website without a clear justification or operator and made up of a few sparse pages, is perhaps the most cited source for legal scholars and other privacy researchers needing to describe cookies. Netscape tapped Whalen to moderate on a couple of its developer forums, where he posted this first cookie explainer with example code. When cookies hit the news, "That's when the world went nuts. It wasn't hard to find a story about how internet sites are 'violating' our privacy with these things called cookies. And, of course, the public didn't know much about it other than to be scared of them. Some folks thought cookies were viruses or worse. Of course, it didn't help that at the time Microsoft was going through trust issues . . . People just thought cookies were another attempt by nefarious developers to steal their information."[166] The Cookie Central site operator asked if he could post Whalen's FAQ, and Whalen spent endless hours responding to new questions and concerns.

It is not surprising that the public was uncertain about cookies or that in February 1996 journalists did not know about DoubleClick's growing advertising network and cross-site tracking. Back in April 1995, when Montulli was explaining cookies on the www-talk list, he stated, "Only hosts in the specified domain can set cookies for a domain. Therefore it is not possible to set a cookie for the B domain unless you are in the B domain."[167] By 1997, the conversation (and participants) had changed and the issue was whether third-party cookies deriving from companies (like those in the advertising business) should be prohibited, or allowed only when the user accepts, or made the default but possible to change. Very often these third-party cookies were set with a request made to fill the image space left by the "img" tag—those tags Berners-Lee disfavored because images were downloaded automatically without the option for intervention like other tags.

In December 1995, a state management subgroup was created within the HTTP Working Group. It was led by Dave Kristol and included Montulli, Amazon's first employee Shel Kaphan, a couple engineers from the W3C group like Holtman, Eric Sink at competitor browser Spyglass Mosaic, and a small group of other developers. The group consulted over email and adopted Netscape's cookie mechanism as a starting point for generating more precise specifications.[168] After the initial draft was published, beginning with version 2.3 in February 1996, "unverified transactions" (later known as third-party cookies) were identified as a significant privacy issue and allowed only if the user could control them through a browser with a default set to reject them.[169] More changes and drafts followed. Last calls for edits were made in the summer, but in October, it was suggested that the specification was too lenient on distinguishing between verified and unverified transactions. After some final adjustments, the specification was published as RFC 2109 in February 1997, as a Proposed Standard called HTTP State Management Mechanism, coauthored by Kristol and Montulli.

As the debate over the use of third-party cookies took off in March 1997 on the www-talk and http-wg lists, Ted Hardie, who worked at NASA's Network Information Center, clarified the specification:

> There is nothing in the spec which forces you back to a "login" method to track unique users. The spec requires that you make sure that the same cookies are not used across domains; it allows you to do what you want *within a domain*. If you want to make sure that a dejanews user sees an advert three times and then gets moved to a different ad, you are welcome to use cookies to do so. If you wish to

use cookies to make sure that a user sees an advert three times *at any site* before moving on to a different ad, sorry. Because the use of the same cookie across sites allows for the creation of very invasive clicktrail analysis, that has been ruled out. There is something which says that if you are using cookies, you need to manage them within your domain.[170]

Hardie explained that cross-site tracking would not be part of the commercial web, that the privacy consumer would not be presented with such a choice, and instead ads would be managed by domain operators.

While the working group set out to resolve some outstanding technical issues, two other unfamiliar voices joined the debate and brought public attention to the normally insular world of the Internet Engineering Task Force (IETF) standards: DoubleClick and the privacy organization Electronic Privacy Information Center (EPIC).[171] In March 1997, Merriman argued on behalf of DoubleClick for modifications to the 2109 specification:

> In its current form, this section of the spec has potentially huge ramifications for Internet advertising networks and remote ad delivery services. Ad networks and remote delivery services use cookies for at least three functions: 1) measuring reach, 2) frequency stop targeting, and 3) user interest targeting. If the new RFC is adopted, these capabilities will be lost for networks, but still available and effective for the largest individual sites which accept advertising. I am unsure if ad networks, which provide an important economic model for smaller web sites, will be able to compete effectively in the long term if this happens.[172]

In April 1997, EPIC sent (and posted) a letter to the IETF, cc'ing Microsoft's president Bill Gates and CTO Nathan Mhyrvold and Netscape's president Jim Barksdale and Andreessen. The letter of support for 2109 concluded by stating, "We believe that 'transparency'—the ability of users to see and exercise control over the disclosure of personally identifiable information—is a critical guideline for the development of sensible privacy practices on the Internet. The alternative would be the surreptitious collection of data without the ability to exercise any control."[173]

The advertising industry also made public appeals, "What concerns us is the tone of the proposal, which is that advertising is not good for us, so we want to avoid it. That begs the question, how is the Web going to be funded?"[174] Advertising representatives also courted Microsoft and Netscape directly.[175] The tangle of technical and social issues resulted in delays, and Kristol decided to find consensus on the technical part; he did so in February 1998, and reintroduced the "unverified transaction" language later.[176] The composition of the steering group had changed in the three years it

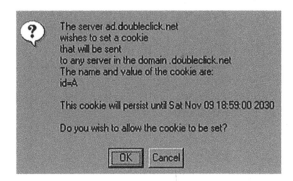

FIGURE 4.1
Netscape version 3.04 cookie warning interface.

had taken to move RFC 2109 along, and the members now wanted more privacy and security.[177]

In the end, Montulli decided not to follow the standard he coauthored; apparently, the drafting committee had gone against his wishes.[178] Both Microsoft and Netscape provided the ability to change preferences to alert consumers before each web server placed a cookie and to reject it (see figure 4.1). Both defaulted to accept all cookies, which went against the 2109 standard but not against the law (see figure 4.2).[179] Montulli, like Behlendorf and others, did not see third-party tracking coming. When it occurred, he assumed that other social forces like laws or government regulation would step in:

> Any company that had the ability to track users across a large section of the web would need to be a large publicly visible company. Cookies could be seen by users so a tracking company can't hide from the public. In this way the public has a natural feedback mechanism to constrain those that would seek to track them ... Governments have an ability to regulate the collection of data by large visible companies and has shown a willingness to do so. The public has a responsibility to keep pressure on both the companies that have the ability to track users and governments to enact reasonable privacy regulations and enforce them.[180]

Montulli insisted that cookies were the most privacy-protective technique available because they provide anonymity and control, and other options for tracking, which would inevitably persist, were harder to observe and disable. He reasoned, "The nature of the advertising business is to collect as much information as it possibly can, so the public needs to push back when it goes too far."[181]

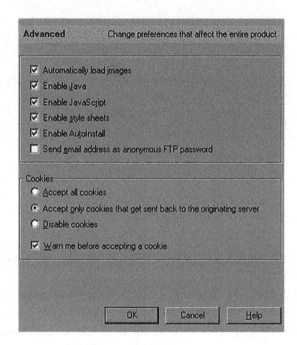

FIGURE 4.2
Netscape version 3.04 cookie settings interface.

Numerous reports were issued by government agencies in the late 1990s that further pushed the rhetoric of consumer privacy and opt-out choice as a self-governance tool. In 1998, the FTC issued a report to Congress called *Privacy Online*, which also misinterprets the *HEW Report* to suit self-regulation, emphasizing consumer notice, choice, access, security, and redress.[182] Further, the agency explained: "In the online environment, choice easily can be exercised by simply clicking a box on the computer screen that indicates a user's decision with respect to the use and/or dissemination of the information being collected. The online environment also presents new possibilities to move beyond the opt-in/opt-out paradigm. For example, consumers could be required to specify their preferences regarding information use before entering a Web site, thus effectively eliminating any need for default rules."[183] The 1998 report found that self-regulation had not been effective on the web and sought to more strenuously encourage development of industry efforts to protect consumer privacy online. By 1999, Pitofsky was pleased to report to Congress that "notable progress" had been made, with more sites making at least one disclosure about their information practices.[184] Yet, although most sites collected information, only a small percentage followed the FTC FIPs. Pitofsky did not stray from industry self-regulation. He concluded his remarks by maintaining that it was not the right time to pass legislation and that more education and outreach was needed, insisting that the "industry must work together with government and consumer groups to educate consumers about privacy protection on the Internet. The ultimate goal of such efforts, together with effective self-regulation, will be heightened consumer acceptance and confidence. [The i]ndustry should also redouble its efforts to develop effective technology to provide consumers with tools they can use to safeguard their own privacy online."[185] A 1998 National Telecommunications and Information Administration (within the DOC) discussion draft titled "Elements of Effective Self-Regulation for Protection of Privacy" suggested using the principles in the *HEW Report*, but focused on consumer transparency, education, and choice.[186] By contrast, in 1998 the Department of Energy's Computer Incident Advisory Capability information bulletin on internet cookies refers to "users" and never mentions the term "consumer." Nonetheless, the bulletin found concerns about cookies unwarranted: "No files are destroyed or compromised by cookies, but if you are concerned about being identified or about having your web browsing traced through the use of a cookie, set your browser to not accept cookies or use one of the new cookie blocking packages."[187]

In 1999, DoubleClick announced plans to merge with Abacus Direct, which triggered several legal proceedings. Abacus marketed consumer-purchasing data to catalog firms engaged in direct marketing and planned to expand to e-commerce retailers. The direct marketing data company earned $11.4 million in revenue of $47 million, while the internet advertising firm lost $18.1 million in revenue of $80 million in 1998. The FTC responded to EPIC's 1999 complaint about the merger by investigating, which prompted DoubleClick to voluntarily put several mechanisms in place to promote consumer choice and privacy education.[188] These voluntary measures were developed by and shared among members of the Network Advertising Initiative (NAI), an industry group established in 2000 by DoubleClick, Engage, 24/7 Media, and the few other web advertising networks operating at the time.[189] Inspired by the DMAA's opt-out system, the NAI developed a "set of self-regulatory standards governing Interest-Based Advertising [and] also pioneered the creation of the industry choice page for consumers."[190] The FTC commissioners unanimously applauded the NAI for developing the self-regulatory system.[191] These had not been the only self-regulatory guidelines available for consideration. In fact, two years earlier, the Better Business Bureau (BBB) announced a more rigorous self-regulatory program that involved verification and dispute resolution services.[192] All served to protect a new form of consumer privacy for the privacy consumer.

In addition to scrutiny from the FTC, ten state attorneys general—led by New York attorney general Eliot Spitzer—began investigating DoubleClick in 2000. The thirty-month-long investigation resulted in a settlement that required websites that partnered with DoubleClick to disclose activities in their sites' privacy policies and required that DoubleClick purchase 100 million banner ads that directed users to its privacy education campaign and opt-out page. DoubleClick was already proactively undertaking these measures due to the foresight and guidance of in-house counsel, the NAI, and a united front of industry associations.

The privacy consumer simply needed more information and better tools to consent by choosing in the marketplace. These same industry associations and multinational companies fought hard to shape the debate similarly in Europe, but on a different front and with very different outcomes. When the American-born privacy consumer of the late 1990s confronted the European anonymous user and data subject, no clear winner emerged.

5 CONTESTING COOKIES FOR DATA PRIVACY

In the first two decades of the 2000s, the convergence of three computer characters—the data subject, the anonymous user, and the privacy consumer—created significant problems for keeping cookies in check. After the dot-com bubble burst, cookies became an issue for "data privacy," a confused and ineffective version of data protection, communication privacy, and consumer privacy.[1] Consent was written into laws in the form of control in data protection, authorization in communication privacy, and choice in consumer protection, but the conflation of characters who would consent resulted in weak and odd articulations of consent across all three areas of law. Over the first decade of the twenty-first century, the US developed and indulged the privacy consumer, who faced off with an increasingly assertive data subject and a less relevant user. A number of failed attempts were made to address a deluge of cookies from a growing number of unknown third parties and increasingly powerful platforms: Users attempted to sue under the Computer Fraud and Abuse Act (CFAA) and Electronic Communications Privacy Act (ECPA); the Federal Trade Commission attempted to modify consent practices for privacy consumers; the European Union attempted to modify consent practices for users; technical standards bodies attempted to create consent mechanisms that would satisfy regulations for privacy consumers, users, and data subjects; and the US and EU attempted to rebuild an international arrangement that was intended to protect data subjects but was housed in consumer protection and taken down with violations of communication privacy. All these attempts failed, not because of the volume of digital information and data collection that people were exposed to daily, but because the ill-fitting laws presented the opportunity for convenient confusion and inconsistency among areas of law that added a layer of politically manifested complexity over the issue of cookies.

COOKIES IN COURT AND ON THE HILL

While state and federal agencies investigated DoubleClick for potential consumer protection violations, numerous class-action lawsuits were filed and consolidated in January 2000 against the company for its use of cookies. Silicon Valley's new geek celebrities were taken to court by the firm of high-profile detractor Bill Lerach, who was known for taking down Enron and called "the most feared man in Silicon Valley."[2] When another of the firm's partners, Melvyn Weiss, filed a class-action claim against Double-Click, Weiss told reporters the company had committed "obvious violations" of federal law and, "we can't even figure out what their defenses are."[3] According to the claim, DoubleClick "covertly, without consent, and in an unauthorized manner implanted Internet 'cookies' upon Web users' computer hard disk drives [and] . . . used those cookies and other data collection methods to secretly intercept and access computer users' personal data and Web browsing habits and has surreptitiously transmitted this personal and private information to DoubleClick for its commercial benefit."[4] The lawsuit alleged that the company had violated the Stored Communications Act, the Wiretap Act, and the Computer Fraud and Abuse Act—that is, user laws.[5]

DoubleClick won. The judge for the Southern District of New York dismissed the federal claims, describing the architecture and technical arrangements using the term "user" colloquially, as an individual operating a computer with authorization who may be identifiable at some but not all levels of interaction. But in its discussion of Title II of the ECPA (the Stored Communications Act), the court included "Web Sites are 'users' under the ECPA" as a heading. Assuming the communication was considered to be held in "electronic storage," the court determined DoubleClick's conduct amounted to an offense under the statute and was trying to determine whether it also met the exception laid out in § 2701(c)(2) for conduct authorized "by a user of that service with respect to a communication of or intended for that user." Working through the definition of "user" in the

ECPA, which recall includes "any person or entity," the court disagreed that "users" need be humans. The court explained that websites also subscribe to ISPs to gain internet access, along with governments, companies, and universities:

> One could imagine a facially sensible argument that Web sites are not "users" of Internet access because they are passive storage receptacles for information; the human is the "user" and the Web site is what is used. However, the Internet's engineering belies this description. Because the Internet functions through packet-switching and dynamic routing, human users do not in any sense connect to a passive receptacle and obtain information. Indeed, no direct connection ever exists between the human user and the Web site. Rather, the human user sends a request to which the Web site must actively respond: processing the request, deciding whether to provide the information sought, obtaining the document from the server, translating the document into TCP/IP protocol, sending the packets and awaiting confirmation of their arrival. Indeed, in a practical sense, Web sites are among the most active "users" of Internet access—their existence and utility depend on it, unlike humans.[6]

The court considered websites to be users under ECPA and that the communication was intended for them, so they were able to consent to or authorize DoubleClick's access. In other words, DoubleClick-affiliated sites consented to DoubleClick's access to the plaintiff users' communications; therefore, no violation of the law occurred.[7] The same reasoning was used to dismiss the Wiretap Act claim: websites were parties to the communication and consented to interception by DoubleClick. The company did not insist its conduct was authorized under the CFAA. Instead, it successfully argued the plaintiffs could not meet the damages threshold of $5,000 required by the statute; after all, the court reasoned, plaintiffs could avoid cookies entirely simply by changing their browser settings or downloading an opt-out cookie from DoubleClick's website, or so the court reasoned.

DoubleClick maintained that none of the statutes applied to its conduct. The defense briefs argued, "In fact, Congress has recently concluded that existing federal statutes 'provide virtually no privacy protection for Internet users.'"[8] The court agreed. Lerach had said, "Nothing makes me happier than crafting a terrific complaint, or a good brief. To me, doing what I'm doing is like a painter painting, or a sculptor sculpting: it's work, but it's a labor of love."[9] But there's only so much an artist can do given their tools.

After the decision to dismiss the federal claims, the parties settled in 2002. DoubleClick agreed to notice practices, a consumer education effort involving 300 million banner ads, and $1.8 million in costs and legal fees. DoubleClick's first chief privacy office, Jules Polonetsky, explained, "The steps we are taking represent by far the most aggressive leadership position regarding privacy within our industry. DoubleClick will continue to provide the same full range of marketing solutions for our clients, buttressed by new and improved internal controls and protections to further safeguard consumer information."[10] Polonetsky joined DoubleClick after serving as the Commissioner of the New York City Department of Consumer Affairs and now runs the nonprofit Future of Privacy Forum, which acts as an important convening force in the privacy community today. Two nonprofit class-action members, EPIC and Junkbusters, however, objected to the settlement, because it did not require the company to do anything it was not already doing.[11] In fact, none of the investigations, lawsuits, or settlements required DoubleClick to do anything beyond what its general counsel and Polonetsky's team had proactively put in place: an aggressive public education campaign that directed consumers to information about data use and provided a means for exercising tracking preferences across the advertising network. Jason Catlett, founder of Junkbusters, was unimpressed. Referring to the consumer protection settlement with the state AGs, he said, "The settlement helps vigilant surfers find out better how they are being observed but doesn't adequately protect the privacy of the vast majority of Internet users."[12]

When Congress probed the legislative landscape in numerous hearings, Catlett testified as a privacy advocate and explained that Junkbusters was a website that people went to in hopes of learning how to stop junk communications like junk mail, junk faxes, junk calls—doing what Senator John McCain called "God's work."[13] The hearings revolved largely around reports issued by the FTC. In 2000, the FTC had issued its third online privacy report.[14] Whereas the two reports in the two previous years had remained faithful to the potential of self-regulatory measures, the 2000 report called for legislation. Catlett argued for more expansive legislation than any member on the panel and beyond the taste of any legislators. Jerry Berman, the Executive Director of the Center for Democracy and Technology (CDT), emphasized the need for federal notice standards to prevent passage of the 200 state bills being debated by state legislators. He explained CDT's mission was "maximizing the democratic potential of cyberspace."[15] Berman argued that users just needed to be empowered to protect their rights.[16] Representatives from the Online Privacy Alliance and AOL, as well as various senators and FTC Commissioner Orson Swindle, argued that the FTC already had plenty of authority to go after bad actors and that self-governance would still work as norms around notice developed. Others argued that the FTC just needed more authority to require privacy policies.

When asked about self-governance, Catlett maintained that the FTC clearly did not have what it needed to stop the junk communications that plagued visitors to his site and that merely addressing notice was a mistake. He argued that the Online Privacy Alliance principles (notice, choice, access, and security), even in combination with technological tools, would achieve little meaningful difference and that the FTC's conclusion that legislation is necessary was unassailable. What kind of legislation? Catlett did not think much of bills that required notice to achieve choice; he wanted consent before any data was collected and other ways for users to protect themselves. He liked the Consumer Privacy Protection Act (CPPA) because it went "beyond the trade issues."[17] Despite its title, the CPPA was expansive.[18] It addressed online privacy, privacy for media consumers, privacy for bankruptcy records, and communications privacy. Dozens of privacy bills were introduced in the 106th Congress (1999–2000) alone. Beyond the CPPA, other bills sought to protect loan applicants, motorists, students, gun owners, bank customers, Wi-Fi users, and social security number holders against entities like law enforcement, advertisers, website operators, and telephone companies. Few of the bills even made it to a committee.

Even though Congress had passed laws specifically for the internet to address platform liability, and having updated privacy laws to protect digital communications, lawmakers were resistant to passing laws that would regulate online data but not offline data.[19] However, FTC chair Pitofsky argued, "The argument that offline and online should be treated in a radically different way just doesn't hold up."[20] Committee reports and panel testimony referenced direct mail databases, banking information, insurance and phone records, and credit card transactions, all of which were certainly stored on computer systems by 2000, but the exceptionalism of the computer did not surface, only the potential exceptionalism of the internet. Congress did manage to pass an internet-specific law for spam emails in 2003. After a growing sense of annoyance and list of state regulations, the federal Controlling the Assault of Non-Solicited Pornography and Marketing (CAN-SPAM) Act to address unsolicited emails in two carefully construed ways. First, the law requires that all commercial emailers provide opt-out options wherein recipients of a commercial message can easily choose to not receive any future emails from the sender. Second, it restricts common and deceptive practices employed by spammers like spoofing the "from" field. The FTC was charged with enforcement. Critics called the CAN-SPAM Act the "You Can Spam Act," because it preempted more restrictive state spam laws, some of which had been declared unconstitutional restrictions on free speech.[21] None of this actually reduced the amount of spam sent, but spam filters built by email providers did the trick.

Meanwhile, Google bought DoubleClick in 2008. Communications scholar Matt Crain describes Google's advertising strategies prior to the purchase as twofold: it presented ads based on the subject's searched on its own search page, and it presented ads through its AdSense program on other websites based on the content of the page, turning the web "into a giant Google billboard."[22] These are considered contextual advertising methods. Feeling the pressure from shareholders, search competitors, and social media companies, Google looked to get into the targeted advertising game. And DoubleClick was on the market. The company had weathered the bubble, downsized, and even managed to start making a profit for the first time. When Google outbid Microsoft in its pursuit of the biggest and best in targeted advertising, Google paid $3.1 billion for DoubleClick, nearly twice what it had for YouTube the year before. Competitors quickly bought up the rest of the advertising players, feeling the same compulsion as Google. Crain notes that buying DoubleClick was a monopoly move by Google that may have represented a lack of choice for the company and signaled the end of choice for everyone else, explaining "[N]early all components of the internet function as terminals of surveillance for platform hubs, sweeping data in and out of the systems of a handful of surveillance advertising's big winners."[23] The merger was a moment that required a public policy response, and groups like EPIC waged a substantial campaign against it.[24] The FTC investigated the merger and privacy advocates pushed for the agency to block it, but the FTC would not and possibly could not consider privacy when it assessed whether the merger would harm consumers.[25] The privacy consumer may have been central to the US approach to internet policy, but they were not accounted for when Google bought DoubleClick.

Although the FTC did not receive additional powers or significant resources to actively pursue privacy beyond the purview of unfair and deceptive practices, nor did it push the boundaries of its antitrust prowess, it did manage to enforce companies' privacy policies to meaningful ends. Starting in the 1990s, the agency accumulated enforcements that mostly resulted in settlement agreements. These agreements are so studied, analyzed, and relied upon that the collection has been called the "new common law of privacy."[26] The Facebook consent decree discussed in the introduction of this book is part of that common law. In it, the FTC took issue with numerous representations Facebook made regarding data settings and data realities. These included changes to the visibility of information that users designated as private and to third-party apps' and advertisers' access to users' data and users' friends' data, and compliance with the US-EU Safe Harbor Framework (discussed in greater detail later in this chapter).[27] The final consent decree, approved in 2012, barred Facebook from making more privacy and security misrepresentations, required the creation of a comprehensive privacy program to assess risks and third-party audits, and demanded express consent before overriding privacy preferences.[28] Privacy consumers were getting somewhere, but not beyond consumer protection, and numerous complaints about violations of the Facebook consent decree went unaddressed.

Users didn't give up though. They continued to seek justice using communication privacy laws, resulting in alternative analyses from other courts that gave mixed reviews of the 2001 DoubleClick class-action case described above. A district court deciding *In re Pharmatrak, Inc. Privacy Litigation* followed the DoubleClick case, but the First Circuit overturned the district court's dismissal, explaining that the lower court had not applied the DoubleClick case properly. Pharmatrak contracted with pharmaceutical companies to provide information about traffic on company websites where people could learn about medications and treatments. Unlike DoubleClick, Pharmatrak services were marketed to the companies as preserving privacy and were not to include the collection of personal information.[29] The First Circuit insisted that consent from the first-party websites needed to be more fully interrogated, because Pharmatrak *had* collected personal information, and it was far from clear that the websites Pharmatrak partnered with consented to such an interception.

User plaintiffs saw hope for new claims when, even as savvy users, they could not avoid the increasingly powerful gaze of what was becoming known as Big Tech, a derogatory term for a handful of dominant global American technology companies (consistently used to refer to Google,

Amazon, Apple, and Facebook). In 2012, reports revealed Google was circumventing users' attempts to block cookies through Apple's Safari browser (opt-in by default) and Microsoft's Internet Explorer (for those who had changed their preferences to opt-out). This set off a flurry of consumer protection investigations at the federal and state level that resulted in over $39 million in settlements and penalties.[30] Users also sued. They argued that Google had intercepted their communications with first-party websites without their consent. Google argued that it was a party to the communications because users (despite cookie-blocking settings) directly sent Google's servers GET requests, asking to retrieve data from a specified URL. As a backup, Google also argued that the code that prompted users' browsers to send the GET request was placed on websites with those sites' consent. The class of users whose browsers sent Google GET requests argued that they were "induced to do so by deceit," but the Third Circuit sided with Google on the first argument and so the backup was not necessary.[31] The court noted the Wiretap Act addresses parties as those who take part in the conversation, which might include parties who have lied or schemed their way into a conversation, like an undercover agent. According to the court, the act does not "exclude intended recipients who procured their entrance to a conversation through a fraud in the inducement, such as, here, by deceiving the plaintiffs' browsers into thinking the cookie-setting entity was a first-party website."[32] The Third Circuit aligned with the Second, Fifth, and Sixth Circuits when it concluded that the users' knowledge and intent are not important to the party exception. Under this interpretation, Google tricked users into communicating with it, and the Wiretap Act does not prohibit tricking people in order to get them to share information.

By and large, legal attempts in the US to contest cookies on behalf of the user and privacy consumer over the course of the first decades of the 2000s failed. Congressional attempts to bolster protections or create data protection for different types of characters also failed.

THE EU COOKIE DIRECTIVE

"A nation gets the level of privacy protections that it demands," Catlett told Senator McCain in 2000, after saying DoubleClick did not set cookies in Germany because of privacy laws. But Europeans were plagued by cookies as well and struggled with many of the same challenges and challengers.

However, these battles were fought on different fronts. As discussed in previous chapters, cookies trigger two separate fundamental rights in Europe: privacy and data protection. This duality is worth repeating and clarifying. Recall, the 1953 Council of Europe's Convention on Human Rights provided a "right to respect for privacy and family life" in article 8 of the convention:

1. Everyone has the right to respect for his private and family life, his home and his correspondence.
2. There shall be no interference by a public authority with the exercise of this right except such as is in accordance with the law and is necessary in a democratic society in the interests of national security, public safety or the economic well-being of the country, for the prevention of disorder or crime, for the protection of health or morals, or for the protection of the rights and freedoms of others.[33]

Data protection was not mentioned specifically. But the 2000 Charter of Fundamental Rights of the European Union protected the right to private and family life in article 7 and also provided a right to the protection of personal data in article 8.[34]

Article 7 states, "Everyone has the right to respect for his or her private and family life, home and communications."[35] Article 8 states:

1. Everyone has the right to the protection of personal data concerning him or her.
2. Such data must be processed fairly for specified purposes and on the basis of the consent of the person concerned or some other legitimate basis laid down by law. Everyone has the right of access to data which has been collected concerning him or her, and the right to have it rectified.
3. Compliance with these rules shall be subject to control by an independent authority.[36]

The five-decade difference between the Council's Convention and the European Union's Charter suggests that privacy is old and data protection is new, but national constitutional rights to privacy and to data protection emerged in parallel. According to legal scholar David Erdos, the communications privacy rights detailed in the previous chapter, as well as those pertaining to the home, are quite old.[37] But he argues a "general right to privacy" in the convention prompted the inclusion of a right to privacy beyond the home and correspondence in national constitutions from the 1960s to the 1990s. These were the same decades in which many European nations added a data protection right to their constitutions. Erdos finds a "close and even symbiotic relationship between these two rights." Navigating the distinction and symbiosis between privacy and data protection was not easy over the first decade of the 2000s, even for Europeans.

The 2002 Privacy and Electronic Communications Directive (ePrivacy Directive), which would eventually become known as the EU Cookie Directive seven years later, was a product of that complicated symbiosis. The ePrivacy Directive was proposed in 1990 as a *lex specialis* (subject-specific law) to "particularise and complement" the Data Protection Directive, a *lex generalis* (general law).[38] When the ePrivacy Directive was finally adopted, it applied to "the processing of personal data in connection with the provision of publicly available electronic communications services in public communications networks in the Community, including public communications networks supporting data collection and identification devices."[39] As explained in chapter 3, it was part of a series of telecommunications directives that required Member States to pass national telecommunication laws to protect both privacy and data on their public networks. Although the 2002 Directive also addressed data breach notifications, the most relevant and prominent goal of the directive was to ensure confidentiality of communication over public networks, including prohibitions on listening, tapping, and storing information without the consent of the users concerned.

Electronic communications services, defined in article 2(c), are about "the conveyance of signals on electronic communication networks" but not on "information society services."[40] This distinction means that the 2002 Directive, applies to telecommunications operators and internet service providers but not to platforms or apps (sometimes referred to as "over the top" services). However, article 5(3), which relates to access to information on a device (such as cookies placed on a user's device for later retrieval), and article 13, which relates to unsolicited communications, apply to *all* entities—and both require user consent and cannot rely on other legal grounds found in the General Data Protection Regulation (GDPR) or elsewhere. As mentioned in chapter 3, the user is defined as "any natural person using a publicly available electronic communications service, for private or business purposes, without necessarily having subscribed to this service."[41] Consent is defined in the ePrivacy Directive by reference to its definition in the Data Protection Directive (and now the GDPR)—symbiosis indeed.

The initial proposal for the ePrivacy Directive from the Commission in 1997 contained no provisions specifically on cookies, but a Dutch Parliament member proposed a prohibition on cookies in the European Parliament's first reading, which stated, "The use of devices to store information or to gain access to information stored in the terminal equipment of a subscriber (such as cookies) should be prohibited unless a prior explicit, well-informed and freely given consent of the users concerned has been obtained."[42] Additionally, the Article 29 Working Party issued recommendations on "Invisible and Automatic Processing of Personal Data on the Internet Performed by Software and Hardware," which stated that the user should be given clear notice and easy tools to exercise the option to accept or reject cookies, but that "browser software should, by default, be configured in such a way that only the minimum amount of information necessary for establishing an Internet connection is processed. Cookies should, by default, not be sent or stored."[43]

Immediately upon seeing the amendment, the Internet Advertising Bureau's (IAB, today the Interactive Advertising Bureau) United Kingdom office launched a "Save the Cookies" campaign.[44] The IAB formed in the US in 1996 and established a vibrant European network within two years.[45] The organization adamantly opposed any type of general opt-in system because the organization feared users would be confronted with multiple pop-up windows seeking prior consent before sending each cookie, thus ruining the online experience for compliant sites and services. The IAB and related industry groups targeted their lobbying efforts at the Commission, national experts, and the Parliament committee rapporteur to change the language before its second reading. The message from the IAB was that the internet could not exist technically or economically without cookies, and after the dot-com bubble burst, the vision of the internet as a site of commercial activity remained fragile. In 2000, the Lisbon European Council had set a bold goal to make Europe the most competitive, knowledge-based economy in the world by 2010, and when considering cookies, policymakers didn't want to drive Europeans away from e-commerce.

Efforts to save the cookies met little resistance from privacy advocates, who were more focused on the campaign to ban spam and governments prioritizing data retention, and the final product (Directive 2002/22/EC, 2002) was a compromise.[46] In the end, article 5(3) of the 2002 ePrivacy Directive states:

Member states shall ensure that the use of electronic communications networks to store information or to gain access to information stored in the terminal of a subscriber or user is only allowed on condition that the subscriber or user concerned is provided with clear and comprehensive information in accordance with Directive 95/46/EC, inter alia about the purposes of the processing, and is offered the *right to refuse* such processing by the data controller. This shall not prevent any technical storage or access for the sole purpose of carrying out or facilitating the transmission of a communication over an electronic communications network, or as strictly necessary in order to provide an information society service explicitly requested by the subscriber or user.[47]

Recital 25 was amended to reflect a "right to refuse" form of consent and the justification for the amendments explained, "Cookies enhance surfing experience and provide for effective web services. Clear and comprehensive information will enable consumers to make an informed choice. In addition, the means to accept and/or reject cookies already exist in most browser software."[48] The UK and its Benelux (Belgium, the Netherlands, and Luxemburg) allies on the Council coordinated a compromise between the Parliamentary Committee (most of whom wanted to restrict Member States from long and loose data retention policies) and the Commission (most of whom wanted to ban spam and saw cookie restrictions as a hindrance to European competitiveness in e-commerce): the Council lifted their opposition to a European-wide ban on spam in exchange for the wording in 5(3) that eliminated "advanced" notice.[49]

Referring to the "victory" that occurred in Europe, one publication stated that opt-in "would have been catastrophic for everything from e-commerce to the set-up of websites, as cookies are used to identify computers and track their users' behavior online."[50] *Computer Weekly* explained, "A ban on cookies would have spelled disaster for Web-based industry. It would have led to the collapse of possibly hundreds of e-commerce companies, hastened the demise of the online advertising industry and stopped companies being able to study usability of a site or improve services for its users by looking at their online behaviour."[51] The responses described a stable infrastructure under threat and in need of care, not an industry with many potential economic models that had just ridden a hype cycle into the dot-com bust.

In 2007, the European Parliament saw another chance.[52] Public pressure mounted that year after the press revealed that Sony had been downloading a Digital Rights Management rootkit (a breed of malicious software that enables privileged access to a computer without detection) to enforce its copyright claim without user consent, and policymakers were driven to revisit the ePrivacy Directive. Parliament seized the opportunity to renegotiate the terms around cookies. In the first reading of the directive, it proposed amendment 128, which read as follows:

> Member States shall ensure that the storing of information, or gaining access to information already stored, in the terminal equipment of a subscriber or user, either directly or indirectly by means of any kind of storage medium, is prohibited unless the subscriber or *user concerned has given his/her prior consent, taking into account that browser settings constitute prior consent, and is provided with clear and comprehensive information* in accordance with Directive 95/46/EC, inter alia about the purposes of the processing, and is offered the right to refuse such processing by data controller.[53]

The Commission rejected this amendment, but in the second reading, the European Parliament tried again. It kept consent in article 5(3) but removed the bit about browser settings:

> Member States shall ensure that the storing of information, or the gaining of access to information already stored, in the terminal equipment of a subscriber or user is only allowed on condition that the subscriber or *user concerned has given his/her consent, having been provided with clear and comprehensive information* in accordance with Directive 95/46/EC, inter alia about the purposes of the processing, and is offered the right to refuse such processing by data controller.[54]

This was either more to the Commission's liking or it went overlooked.

Before the final signing of the amended directive, thirteen Member States of the Council joined in an Addendum to clarify how cookie consent was to work in the new version of the ePrivacy Directive. They declared, "As indicated in recital 66, amended Article 5(3) is not intended to alter the existing requirement that such consent be exercised as a right to refuse the use of cookies or similar technologies used for legitimate purposes."[55] Recital 66 of the new version of the ePrivacy Directive maintained the old language about the importance of providing users with clear and comprehensive information and offering a right to opt-out, ending, "Where it is technically possible and effective, in accordance with the relevant provisions of [the Data Protection] Directive, the user's consent to processing may be expressed by using the appropriate settings of a browser or other application."[56] With this interpretation, the newly introduced "consent" in 5(3) was a right to refuse cookies, which was no change at all.

However, the Article 29 Working Party released an opinion on online behavioral advertising in 2010 that broke 5(3) down into clear notice and effective consent, which it further interpreted in 2011, in accordance with the Data Protection Directive, to mean "prior" consent.[57] National laws implementing what became known as the Cookie Directive thus varied.[58] In fact, the Article 29 Working Party drafted guidance every year for four years on cookies and consent.[59] Yet, national legislation on consent and browser settings remains rife with variation.[60] These inconsistencies were seen as a problem for the Single Market and increasingly for the fundamental rights of Europeans, but the ad tech industry and privacy professionals moved forward smoothly with four main takeaways from the 2009 update: 1) cookie notices should be prominently displayed; 2) action, like remaining on a site, amounts to consent; 3) users must be given means for controlling cookies, except those designated necessary; and 4) clear and comprehensive information about cookies in a cookie policy must be available.[61] The changes increased the prominence of cookie notifications in the form of banners, pop-ups, and separate cookie policies, but didn't do much otherwise.

BUILDING DATA PRIVACY FOR COOKIES

By the mid-2000s, policymakers operated with growing confidence derived from privacy research, tools, and professionalization. Privacy researchers, privacy engineers, and privacy professionals all rapidly proliferated over the period and provided different opportunities for understanding and governing cookies, but few were interested in parsing their work into contributions to specific areas of law like consumer protection, communication privacy, or data protection. Because the FTC was the de facto audience for policy solutions in the US, these efforts were funneled into or produced with an eye toward relevance within consumer protection, even if privacy research, tools, and professionalism developed to understand the practices of different populations of data subjects or technical standards for users.

Early cyberlaw was somewhat dominated by legal and economic scholars, and legal scholars with computer science know-how, who were drawn to an exciting, seemingly new frontier often described as a Wild West that could provide a more perfect market or that was ripe for technical solutions. The novelty of internet issues confronted existing critical, feminist, and philosophical work from privacy scholars like Anita Allen, Julie Cohen, and Helen Nissenbaum.[62] Critical of privacy as control and individual choice, these projects pointed out conceptual limits and were bolstered by empirical evidence.

Much of the empirical evidence came from the expanding ranks of scholars attracted to the issues under the umbrella of "privacy." This group used social science methods to produce important work that powerfully shaped policy debates. Their works were and are referenced regularly in policy workshops around the world, and the researchers are leaders and participants in policy processes. Behavioral economist Alessandro Acquisti and his colleagues' research out of Carnegie Mellon's Heinz College of Information Systems and Public Policy led the economic analysis of privacy using behavioral experiments to understand consumer privacy and decision-making.[63] Joseph Turow's team at the University of Pennsylvania's Annenberg School of Communication produced surveys that repeatedly revealed that people misunderstood the presence of a privacy policy as a signal of privacy protection.[64] danah boyd, the coder who helped build GRAMMATRON, undertook groundbreaking anthropological work on kids' use of social media that conflicted with the flippant assessment of young people's disregard for privacy.[65]

Others wanted to build new tools and systems to promote or ensure privacy. Part of the privacy by design movement, privacy engineering began as a set of privacy-enhancing technology (PET) projects. Lorrie Cranor, an early leader in the PETs movement, transformed the projects into a field known today as privacy engineering and trained a new generation of privacy engineers.[66] One such engineer is Aleecia McDonald, who cochaired the WC3's Tracking Protection Working Group (TPWG), which was an effort to establish international standards for a global Do Not Track (DNT) consent mechanism. The TPWG and DNT saga are infamous among those who work in privacy. If you look closely at DNT, it's clear that cookie controversies were never just about managing information overflow to help people make accurate choices about sharing their data online; they were about politics and investment in the status quo of computing arrangements.

The DNT saga began in 2007, when the World Privacy Forum led a group of privacy advocates to propose the idea of a DNT list. Inspired by the popular Do Not Call Registry, which is a list of phone numbers from people who do not want to be contacted by telemarketers, the DNT list would have also been maintained by the FTC. Sites and ad networks that placed persistent cookies would populate the FTC list by registering cookie-placing domains, and then browser plug-ins could be developed to block those domains.[67] The networks would have been charged with keeping the list up to date, and the FTC with promoting it. Microsoft took up the idea. Internet Explorer 9 allowed third parties to create blocking lists that users could add to the browser. Tracking Protection Lists carried two main benefits: they weren't created by Microsoft, keeping the company away from antitrust issues, and they were enforceable—content from the domains on the list simply would not load. However, privacy lists could be quite ineffective. For example, TRUSTe, a privacy compliance technology company, distributed one of the first lists, which included 4,000 domains on an allow list and none on the block list.

In 2009, Google offered an opt-out option in the form of a browser add-on for DoubleClick. Recall that users could already opt out of cookies set by members of the NAI, but this system used opt-out cookies that could be delivered when a user went to each member's website or from the NAI website. The opt-out cookies were deleted every time a user went into their browser settings and deleted all cookies. Google's add-on stabilized the DoubleClick opt-out cookie, making it possible to delete all cookies without eliminating the DoubleClick opt-out cookie. In discussing the flaws of opt-out cookies, privacy researchers Chris Soghoian and Dan Kaminsky surfaced the idea of a browser header that could be added to Soghoian's Targeted Advertising Cookie Opt-Out (TACO) add-on.[68] Each HTTP request would be sent with a header that signaled the user had opted out of cookie tracking.

Although the header approach (changing fields used to pass information between client and server sent with each HTTP request and response—like Set-Cookie used to check for cookies previously sent by the server discussed in chapter 4) found little support among advertising networks, it found an audience with the privacy community and with the FTC, where Soghoian went to work in 2010. FTC chair Jon Leibowitz showed support for the DNT header during his Senate testimony before the Committee on Commerce, Science, and Transportation in July 2010. He specifically noted it would make it easier for consumers to opt out, instead of having to make choices on a site-by-site basis. Senator Jay Rockefeller IV, who chaired the committee, asked for that opt-out. In Rockefeller's opening remarks for the hearing, he questioned whether consumers read the fine print and whether consumers could stop tracking even if they knew about it. He emphasized,

"The *consumer* I'm concerned about is not a savvy computer whiz kid. I'm not talking about a lawyer who reads legalese for a living and can delve into fine print of what privacy protections he or she is getting. I'm talking about ordinary internet users."[69] He described an ordinary internet user/consumer (alternating between the two terms) as a fifty-five-year-old coal miner sending an email to his son at college, a thirty-year-old mother looking up the best place to take her sick toddler, and a sixty-five-year-old grandfather looking to view family pictures and reconnect with old friends. He stressed that all these users/consumers should be able to prevent tracking. FCC chair Julius Genachowski also testified on the panel but offered little beyond support. He spoke about how the FCC partnered with the FTC to meet the needs of consumers: "It is important that uncertainties in the regulatory framework be resolved. What matters most is the consumer."[70] Everyone seemed to agree: what was needed was an easy opt-out mechanism for privacy consumers.

When the World Wide Web Consortium (W3C) took up the tracking issue, Microsoft and advocacy groups still preferred DNT lists because headers were only requests that could be ignored, while advertising groups preferred Google's opt-out cookies.[71] Soghoian had collaborated with Mozilla engineer Sid Stamm to build a prototype Firefox add-on with two HTTP requests, one that signaled opting out of behavioral ads and one that opted out of tracking (the latter applying more broadly). Stamm had advanced the DNT header in the Firefox browser, with a simple header and interface by early 2011. Shortly after, Stamm, along with privacy researchers Arvind Narayanan and Jonathan Mayer, submitted a proposal for a standard DNT header to IETF, but the organization declined. The W3C, on the other hand, launched the Privacy Interest Group (PING) to consider privacy across all web specifications. Participants rallied for a lengthy workshop and began meeting as the Tracking Protection Working Group (TPWG).[72]

Participants from tech companies, privacy and tech nonprofits, and advertising associations needed to determine how a server responded to a DNT header request—and it got ugly. In her recounting of the DNT standards process, TPWG cochair McDonald shares highlights of "less than good faith" actions and the pugnacious nature of the group.[73] She describes the TPWG email interactions that quickly ignited "into flame wars" and how participants disparaged others in the press, "including personal attacks and name calling."[74] They complained about procedure and used stall tactics, "giving speeches" and "grandstanding."[75] These were minor and common compared to the "derailment of careers."[76] Internet researcher Nick Doty's analysis of the DNT standard uses organizational headings like "toxicity and personal attacks," "difficult people," "animosity," and "bad faith." Doty's interviews with participants reveal DNT was so contentious in part because it was too widely used and therefore too disruptive. When Microsoft turned DNT on by default in 2012, the number of browsers sending DNT signals was far more than many participants bargained for. As one participant stated, "What is your profit margin? If your profit margin is smaller than the DNT adoption rate, then you must ignore all DNT signals or you'll go out of business."[77]

Microsoft's decision to change all its browsers to opt-in might sound reasonable, but McDonald refers to it as "Microsoft's Chaos Monkey Moment."[78] The TPWG had already agreed on two points. First, users on general purpose browsers like Explorer would have to select either DNT or something like "tracking is fine." There would be no default for these browsers (see figure 5.1). Although users would make a selection, there were doubts that users were really getting what they were trying to choose. McDonald worried that DNT "would inherently be a deceptive practice, since despite a catchy name it seemed unlikely to actually stop tracking."[79] Indeed, research showed that users expected that if they selected Do Not Track, they would not be tracked; however, the signal wouldn't prevent all data collection, so the terms of what would still be collected had to be negotiated.[80]

Second, older browsers with no DNT settings would send a default signal, either track or not track depending on whether the browser was in the US or EU. For older browsers in the US that sent no signal, users would have an opt-out default. For old EU browsers, no signal meant users had not consented and thus could not be tracked. This is particularly interesting because in 2012, the European Commission sent a letter to the W3C expressing little fuss over an opt-out default and instead insisted that users should be prompted to make a choice when they downloaded or updated their browser: "It is not the Commission's understanding that user agents' factory or default setting necessarily determine or distort owner choice."[81] But the TPWG took the Article 29 Working Party's guidance, which demanded prior explicit consent, on the changes to the ePrivacy Directive seriously.[82] The position defied protests from the IAB. The IAB vice president expressed

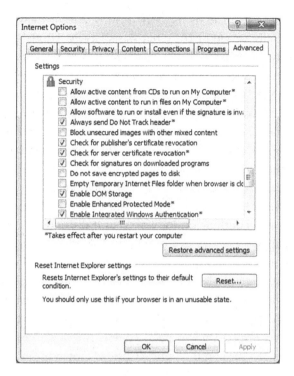

FIGURE 5.1
Internet Explorer version 10 Do Not Track settings interface.

numerous objections, "I have serious problems with the way this group works and operates. I do not believe that we need to delve into (European) legal discussions . . . [Y]ou are pushing so hard for the acceptance of Article 29 WP opinion as the word of God on data protection issues . . . and I don't understand what you are trying to achieve with this. We may like what Article 29 WP says or not, but FACT is that it is JUST an opinion. It is not the law. And, frankly the UK, one of the most engaged EU Member States, is not following the supposed 'baseline.'"[83] Others in the TPWG worried that the specification would be ignored if it didn't meet European legal standards and some didn't understand how "prior consent" could be interpreted other than an opt-in default.[84]

McDonald explained that some participants' involvement in the W3C effort was motivated by fear of increased legislation around the issue of cookies. On the state level, California's Senator Alan Lowenthal proposed drafting statewide regulations for DNT in early 2011.[85] California Assembly Bill 370 finally passed in 2013, but was a largely ignored transparency law that merely required California-based companies to tell users how they respond to DNT requests.[86] Over the course of 2011, Congresspeople proposed the Do Not Track Me Online Act, Commercial Privacy Bill of Rights Act, Consumer Privacy Protection Act, Do-Not-Track Online Act, and Do Not Track Kids Act.[87] Privacy scholars and researchers overwhelmingly agreed that none of them were sufficient.[88] Efforts at the federal level failed entirely. Without clear legal support, the DNT standard was largely ignored.

Legal pressure in the US mounted but fizzled, much the same way as DNT did in W3C. Even with all the ire and pandemonium, the working group managed to continue on for years, but it disbanded in 2019, citing insufficient deployment and lack of support.[89] As of summer 2022, if you run a Google search in a Chrome browser for how to turn the DNT signal on in Chrome, the Google help page explains, "Most websites and web services, including Google's, don't change their behavior when they receive a Do Not Track request. Chrome doesn't provide details of which websites and web services respect Do Not Track requests and how websites interpret them."[90] A user can turn on the Google Chrome browser's DNT setting, but Google sites will not respect the preference.

THE TRANSATLANTIC BREAKDOWN

The wheels really started to come off in 2015, when Safe Harbor fell apart. Constructed between 1998 and 2000 to accommodate the reality that US law would not meet the adequacy standard required to transfer data outside the European Union, the Safe Harbor Privacy Principles were a patch that held things together between the US and EU for a remarkably long time. Under the EU Data Protection Directive, data couldn't be sent out of the EU unless it was to a country with laws that provided an adequate level of protection, the data subject agreed to the transfer, or standard corporate clauses had been authorized. Safe Harbor was a workaround. US companies self-certified to the US Department of Commerce that they followed seven principles. In 2000, the European Commission declared companies that had self-certified met EU data protection requirements to transfer data.

When Edward Snowden, an IT analyst working as an intelligence contractor, disclosed to the world the invasive arrangements between numerous dominant US tech companies and US intelligence services in 2013, Safe Harbor was in trouble. One of Snowden's many revelations was the Prism program, which involved companies like Google, Microsoft, and Yahoo! collecting internet communication on behalf of the US National Security Agency (NSA). Demands for data through the Prism program were conducted under section 702 of the Foreign Intelligence Surveillance Amendments Act (FISA) of 2008 for surveillance inside the US and Executive Order 12333 for surveillance outside the US. FISA is the US communication privacy law that prescribes procedures for accessing foreign surveillance inside the US with the compelled assistance of service providers, while EO 12333 governs surveillance outside the US conducted by the NSA that involves exploiting the vulnerabilities of telecommunications infrastructure.[91] In 2005, the *New York Times* uncovered the Terrorist Surveillance Program (TSP).[92] Since 2002, after the 9/11 terrorist attack, the NSA had targeted domestic email and telephone communications. FISA required a court order, obtained by showing probable cause to believe that the target of the surveillance was an agent of a foreign power, prior to or within seventy-two hours of initiating a wiretap.[93] Through the TSP, the Bush administration had been skipping this step.

Dozens of telecommunications companies were sued. Telecom service providers are only to provide assistance to intelligence agencies if they receive either a court order or certification that no warrant or court order is required by law, that all statutory requirements have been met, and that their specific assistance is required.[94] In early 2006, the Electronic Frontier Foundation (EFF) and others filed class-action lawsuits, forty of which were consolidated and transferred to the Northern District Court of California, accusing the "telecom giant of violating the law and the privacy of its customers by collaborating with the NSA in its massive, illegal program to wiretap and data-mine Americans' communications."[95] Because the program captured phone communications and emails, the claims referred to subscribers and users. The claims used different terms to express expectations and cultural practices around these distinct technologies, but they were packaged together in the legal claim for violations of communication privacy law.

While Congress quickly began steps to amend FISA in light of the controversy surrounding the TSP, it debated granting retroactive immunity for companies that had complied with it and were being sued by users.[96] The issue was central to a push from the government to reconsider and redefine internet privacy. The Principal Deputy Director of National Intelligence, Donald Kerr, explained at the time: "Protecting anonymity isn't a fight that can be won. Anyone that's typed in their name on Google understands that . . . Our job now is to engage in a productive debate, which focuses on privacy as a component of appropriate levels of security and public safety . . . I think all of us have to really take stock of what we already are willing to give up, in terms of anonymity, but (also) what safeguards we want in place to be sure that giving that doesn't empty our bank account or do something equally bad elsewhere."[97] In a letter imploring House Speaker Nancy Pelosi to support the bill, which included blanket immunity for companies that had participated in the TSP, the attorney general and director of the NSA emphasized that any amendment must provide an effective immunity provision. The Senate Select Committee on Intelligence had made it clear that the intelligence community could not get the information it needed without service providers, and the attorney general and NSA director worried companies would not help if they faced the threat of private lawsuits every time someone believed they provided assistance.[98] The White House had already declared it would veto any bill that passed that did not include a blanket retroactive immunity provision for these companies.

But the EFF insisted that anonymity had been fundamental to America's founding, citing the Federalist Papers, and could not be set aside. A liberal "netroots" movement arose to support the lawmakers who were pushing to eliminate or limit the immunity. Political blog Firedoglake planned a "money bomb" protest—a one-day push to raise money for a candidate voting a particular way to create buzz around the issue—to support Rand Paul.[99] On then-senator Barack Obama's social media site, my.barackobama.com, the Get FISA Right Group became the site's largest group, with 14,500 Obama supporters trying to convince him to vote to remove the immunity for companies that had helped with the TSP.[100] He and sixty-six other senators voted to keep it.[101] In July 2008, Congress enacted section 702 of the FISA Amendment Act, which, in addition to providing telecommunications companies immunity, broadened the scope of FISA to allow foreign intelligence surveillance in a bulk fashion, without going through procedural hassles for individual targets.

Far from protecting transfers of foreign data, the FISA amendments reinforced surveillance of communications "outside" the United States, by relying on the increasingly less useful distinction between foreign and domestic data. When Kerr argued for a reconceptualization of privacy after the TSP became public, he insisted that foreign intelligence gathering had to be done on American soil because, as the US came to dominate global telecom, a growing amount of foreign communication was moved through American channels. Yet when Obama administration officials defended the Prism program after Snowden made it public, they emphasized that the targets were *outside* the US—people like Max Schrems.

Although the Austrian Max Schrems was only a law student when Edward Snowden blew his whistle, he filed a series of complaints with data protection offices across Europe arguing that the companies had transferred his data from the European servers to their US servers, where the data was not adequately being protected. Schrems drafted the complaint on behalf of Facebook users, enforcing the legal rights of the data subject and making no mention of consumers.[102] In considering Schrems's claims against Facebook in Ireland, the director of the Irish Data Protection Commission determined that the company had operated within the Safe Harbor agreement, but Schrems appealed, and the High Court of Ireland referred the issue to the European Court of Justice (ECJ). The court invalidated Safe Harbor. It ruled that national data protection authorities have a responsibility to investigate data protection violations, but that only the ECJ could invalidate an adequacy determination, which it did (see figure 5.2).[103]

The ruling shook the tech sector and its lawyers. "Aside from taking an ax to the undersea fiber optic cables connecting Europe to the United States, it's hard to imagine a more disruptive action to transatlantic digital commerce," wrote the trade organization Information Technology & Innovation Foundation.[104] This quote was read at the opening of a joint congressional hearing to emphasize the magnitude of the court's decision and the urgency of coming up with a new agreement.[105] The congresspeople repeated the benefits of the agreement to Europe. They noted that EU Member States were given discretion to balance national security needs with data privacy rights, and that the Court of Justice did not undertake a rigorous factfinding mission when making its decision, but the US seemed much more eager than the EU to establish a new arrangement.[106] The Chamber of Commerce, the FTC, the Department of Commerce, the European Commission, and representatives from European Member States worked to shore up the hole that had punctured international data flows

FIGURE 5.2
Edward Snowden's tweet congratulating Max Schrems after the ruling that invalidated the EU–US Safe Harbor Agreement.

before the enforcement moratorium declared by the Article 29 Working Party ran out at the beginning of 2016. In February 2016, the new Privacy Shield was announced. It contained three components: 1) US companies transferring data from Europe had to meet obligations regarding data processing that were monitored by the Department of Commerce and the FTC; 2) Access to such data by public authorities for national security reasons would be subject to clear limitations, safeguards, and oversight; 3) EU citizens who suspected misuse were entitled to several forms of redress directed through the company; EU data protection agencies could go to the Department of Commerce and FTC, and intelligence complaints could be directed to a new ombudsperson created within the US State Department. In July 2016, the European Commission deemed it adequate for transatlantic data transfers under EU law.[107]

Schrems immediately filed complaints again. In July 2020, the European Court of Justice invalidated the European Commission's adequacy decision, ending the Privacy Shield's legitimacy (though the US Department of Commerce still administers the program).[108] Even though additional protections for those outside the US had been put in place by President Obama shortly after the Snowden revelations and by Congress through the Judicial Redress

Act in 2017, the ECJ invalidated the Privacy Shield just as easily as Safe Harbor.[109] The second ruling was not a surprise. Members of the European Parliament had voiced their disappointment in the Privacy Shield, criticizing rules that had since been issued to allow the NSA to share data with other agencies; the numerous vacancies on the oversight board; the status of the ombudsperson, which alternated between a Silicon Valley executive and an absent seat; and the fact that only a fraction of US companies even bothered to join.[110]

Europeans were also disappointed that the US Congress and Trump administration reversed the first significant steps that the FCC had taken to protect user privacy. Since the 1990s, the FCC had struggled to secure even meager protections for customer data. In 1998, the FCC promulgated regulations implementing the Telecommunications Act of 1934 to limit customer proprietary network information (CPNI). CPNI included data about network usage, equipment, levels and types of service, and numbers called, and by the 1990s, it was being used to more effectively market services or technology to customers. Congress enacted section 222 of the 1996 Telecommunications Act, which treated CPNI as confidential and required getting customer "approval," but didn't define approval. When the FCC implemented rules for "approval" to mean an opt-in consent that could be oral or electronic, it was challenged by many companies as an unconstitutional restriction on their ability to engage in commercial speech with customers. The regulation was struck down by the Tenth Circuit. The court found that the FCC failed to articulate the privacy interests, which needed to be substantial enough to justify limiting the speech of companies.[111]

In 2016, FCC chair Tom Wheeler initiated a proposal to give consumers "choice, transparency, and security" in the information used and shared by ISPs.[112] The year before, the FCC reclassified broadband internet access from an information service to a telecommunications service, thereby triggering the process to establish privacy rules for internet providers that the FTC would no longer have jurisdiction over. The proposal would have required opt-out options for data used to market other communication-related services, even with their affiliates, but an opt-in system for "sensitive information," like location, children's information, and apps used. The requirements didn't go over well with ISPs or the Republican commissioners, who argued it was corporate favoritism. One Comcast executive wrote that the proposal was "inexplicably targeted to block ISPs . . . from entering and competing as disruptors and upstarts in the online advertising marketplace [currently] dominated by edge providers and other non-ISPs."[113] Wheeler insisted that ISPs access data to provide internet connectivity, not because a consumer has chosen to visit their website or use their particular service, and that the FCC was not prohibiting the use of the data, just requiring explicit consent.[114] When FCC chair Ajit Pai took over at the commission in 2017, he announced his intention to delay the privacy rules that had been adopted the year before, citing discordance with the FTC's approach and objections from the two Democratic commissioners.[115] In 2018, Congress stepped in and passed the Congressional Review Act, which repealed the privacy rules and prohibited "substantially similar" rules from being reinstated, and the FCC reclassified ISPs as information services, sending user privacy back to the FTC.

Meanwhile, the EU moved forward with plans to dramatically reinforce protections for data subjects. The second decade of the century was a pivotal moment in EU history, and most people felt it through data protection. In 2009, the Lisbon Treaty transformed the human rights landscape. The 2000 Charter of Fundamental Rights of the European Union, within which privacy and data protection are given their own distinct articles laying each out as unique fundamental rights (discussed at the beginning of this chapter), was originally a political document intended to recognize a synthesized set of national and international obligations. But this political synthesis became legally binding in 2009 under the Treaty of Lisbon, which gave the EU a legal basis for comprehensive legislation across the Community. Prior to the agreement, the EU was limited to the legal basis of the internal market, which only justified directing national laws to approximate one another so as not to inhibit the free flow of data across borders. The Cookie Directive is a relic of the old limited legal basis. After legally binding Member States to the Charter, the EU could pass laws furthering the rights therein. And that's what it did. In January 2012, the EU Commission published its proposal for comprehensive data protection legislation for the entire Community and announced plans to next pass ePrivacy laws to match.

With the announcement of the General Data Protection Regulation (GDPR), the EU launched an initiative to draft legislation protecting the fundamental rights of data subjects through consistent, holistic rights matched by significant obligations on those who process or control data.[116]

With the threat of increased fines up to 4 percent of total yearly revenue, the hard-hitting new rules were adopted in 2016 and took effect in May 2018. The GDPR changed consent as a legal basis for processing personal data, making it a much less attractive option from the list of six, which also included the processing to perform a contract, for a legitimate reason, vital reason, or in the public interest, and to fulfill a legal requirement, still intact from the 1995 Data Protection Directive. The GDPR succeeded in changing the meaning of consent, but the definition in article 4(11) remains largely the same: "'Consent' of the data subject means any freely given, specific, informed and unambiguous indication of the data subject's wishes by which he or she, by a statement or by a clear affirmative action, signifies agreement to the processing of personal data relating to him or her." The power of consent is further reinforced by article 7, which requires the data controller to demonstrate a proof of consent, establishes a right to withdraw consent (further expanded upon in recital 42), and states that consent must be clearly distinguishable from other terms. Recital 32 removes the possibility of opt-out consent by prohibiting silence, inactivity, and pre-ticked boxes as sufficient evidence of affirmative consent.

Recital 32 also states that the data subject may "choos[e] technical settings for information society services" as a means of consent, but the specificity requirement in the definition of consent creates a challenge for existing approaches to browser settings. The Do Not Track header, for instance, may not meet this requirement. The 2017 guidance from the Article 29 Working Party mentions browsers only once, briefly explaining, "An often-mentioned example to do this in the online context is to obtain consent of Internet users via their browser settings. Such settings should be developed in line with the conditions for valid consent in the GDPR, as for instance that the consent shall be granular for each of the envisaged purposes and that the information to be provided, should name the controllers."[117] The specifics of browser settings and cookies were to be addressed in the ePrivacy Regulation, proposed in January 2017 to harmonize Member States in their implementation and alignment with GDPR. Because the EU Cookie Directive derives its definition of consent from the EU data protection law, the GDPR's shift to explicit, prior consent unequivocally changed the default for cookies to opt-in and requires specific consent. The change drove cookies into retirement.

6 CONCLUSION: THE (FORCED) RETIREMENT OF COOKIES

At the beginning of 2020, Google announced it would retire the third-party cookie. In a company blog post titled, "Building a more private web: A path towards making third party cookies obsolete," Google announced a pivot.[1] Although other browsers like Firefox and Safari were already blocking third-party cookies by default, Google's decision to phase out cookies and replace them with something the company had yet to develop was a big deal. At the time, Google Chrome held about 65 percent of the browser market worldwide, 50 percent in the US and 60 percent in the EU, and its decision to restructure the web to rely on first-party data collection in some vague, speculative way meant shaking up arrangements that had been lucrative for some and ruinous for others over the two previous decades.[2]

Google and other surveillance capitalism companies have not given up on data-driven, ad-supported services.[3] They are working hard to develop new advertising models that will continue to convince companies, causes, and campaigns to allocate advertising dollars to their platforms over, say, podcasts or billboards or publisher sites. The retirement of cookies is an end and a beginning, cause and effect. The cookie era is ending because of these shifts, pushed forward by equal justice efforts, antitrust scrutiny, mental health concerns, democratic fears, and political shifts. The end of the era triggers reimaginings across parties like advertisers, political incumbents, publishers, tech companies, and the public. Efforts to remodel surveillance capitalism and targeted advertising are, not coincidentally, taking place amid a much larger bipartisan techlash, which includes a number of voices fighting to revise expectations for the future of technology and society. Such revisions attempt to reimagine the networks and governance structures—the sociotechnical reimaginaries—after a series of cultural shifts, from high-profile Russian bots targeting American voters to increasing concerns about biased algorithms impacting social justice and mental health on addictive services.

Each computer character provides an opportunity for intervention and world building in these sociotechnical reimaginaries. The US urgently needs to create a data protection regime oriented around a distinct data subject that captures the power dynamics of the information economy and the subject access rights that serve as a backstop to vulnerable internal measures. User privacy presents a possible path for consent that would represent a renewed effort in the history of computing to recenter the democratic stakes of the debate. The privacy consumer deserves a higher consent standard than consumer protection may be able to provide. These computer characters are powerful but not enough. New characters must also be developed to fight for meaningful governance of technology.

BEYOND CONSENT THEATER

When Google announced the retirement of third-party cookies, it was not playing the role of the revolutionary. The business model that made some American tech companies the richest in the world would not work under an arrangement based on affirmative, prior consent—the kind demanded by GDPR. Under GDPR, even first-party cookies (and other tracking mechanisms, of course) require consent, unless they are "strictly necessary." These are trackers necessary for the site to function (e.g., an e-commerce site's ability to hold items in a shopping cart) and are often session cookies that expire after a short period of time or when the browser is closed. But nonessential trackers, such as those used to store passwords or language preferences, require consent, as do analytic trackers and marketing trackers.

Although the GDPR is a regulation (with almost no room for national variation), EU Member States still had to update their domestic communication privacy laws (based on the ePrivacy Directive) to comply with the GDPR. Then, national data protection agencies had to figure out how to interpret both the GDPR and their domestic communication privacy law for enforcement purposes. In Germany, the Telemedia Act had not yet been updated to prohibit opt-out consent for cookies and caused some inconsistent rulings in lower courts, so the German Federal Court of Justice sought clarity from the European Union Court of Justice (ECJ).[4] On October 1, 2019, the ECJ handed down a judgment in the *Planet49* case, which provided guidance on pre-checked boxes for cookie placement, consideration of whether the personal nature of the data matters, and requirements for the duration of cookies in the notification. In interpreting article 5(3) of the ePrivacy Directive, the ECJ concluded that pre-checked boxes do not constitute valid consent, expiration dates on cookies should be disclosed, and different purposes could not be bundled into the same consent request—and the court confirmed that these rules apply to cookies irrespective of whether the data is personal or not.[5]

The French CNIL's enforcement brought even more clarity. It took two years, but the CNIL replaced its cookie guidelines in 2020.[6] The guidelines repealed the agency's previous position, which accepted that scrolling down or remaining on a site or service met the standard for consent. The new guidelines stated clearly, with extensive complementary examples, that consent must be "unambiguous."[7] The absence of active and specific consent must be interpreted as a refusal. Most importantly, the CNIL highlights that the means of consent must be as easy as the means of refusal. In January 2021, the CNIL sanctioned Google €150 million and Facebook €60 million for non-compliance with French law. The agency announced, "[T]he websites facebook.com, google.fr and youtube.com offer a button allowing cookies to be accepted immediately. On the other hand, they do not put in place an equivalent solution (button or other) to allow the Internet user to easily refuse the deposit of these cookies. Several clicks are necessary to refuse all cookies, against only one to accept them."[8]

The Belgian data protection agency went after IAB Europe for its Transparency and Consent Framework (TCF). TCF includes a list of vendors, a list of consent management providers, and a set of technical specifications developed by IAB Europe to meet the GDPR consent standards while allowing for real time bidding on advertising space in order to deliver targeted ads. When an individual visits a site for the first time, a consent management system presents preferences that, when selected, are sent along with the other user data shared with the ad vendors. Most European websites use a consent management system, but the Belgian Data Protection Authority found consent obtained through the TCF system invalid. The agency reasoned that TCF sets up numerous recipients of consent, and thus the user would need too much time to read and investigate notices to be sufficiently informed and would have a much harder time withdrawing consent than giving it.[9] Between the ECJ and aggressive data protection agencies, GDPR has been interpreted to largely end consent theater.[10] In Europe, affirmative, prior consent is now required for data collection, and simply remaining on a site or service would not constitute consent. In fact, it must be just as easy to consent to data collection as it is to reject it.

Another consent success can be found in Illinois. In 2008, the state unanimously passed the Biometric Information Privacy Act (BIPA), which requires a person's written consent after being informed in writing about what biometric data is being collected and stored and for what purposes.[11] BIPA is a powerful tool in a moment when the bias of algorithms has become a mainstream topic of conversation, thanks in large part to African American and feminist technology scholars and activists. In 2020, the Duke and Duchess of Sussex, Prince Harry and Meghan Markle, interviewed internet studies scholar Safiya Noble about how systemic oppression is built into technical systems.[12] On the comedy show *Full Frontal with Samantha Bee*, Noble explained how facial recognition and AI are designed and wielded by dominant groups targeting the disenfranchised.[13] BIPA has gone mainstream too, inspiring political pressure to regulate the collection and use of biometric data in the private and public sector.[14] Several high-profile companies' violations of BIPA spurred more awareness of the issue. Facebook photo tagging, Google Photos, Apple Photos, TikTok, and Snapchat smart filters all triggered BIPA lawsuits. In the middle of these lawsuits in 2016, the original sponsor of the bill, state Senator Terry Link, tried to quietly weaken BIPA, allegedly at the behest of Silicon Valley lobbyists, but pulled his amendments after receiving significant attention.[15] Facebook settled its BIPA class action for a $650 million price tag and an agreement to follow direction on how to obtain opt-in consent.[16] Google settled for $100 million, TikTok for $92 million, and the others are still being litigated. In May 2022, Clearview AI, a small, secretive company that controversially sells access to its facial recognition databases built on billions of photos scanned from social media sites and the rest of the web—a service deemed illegal in much of Europe and Canada—lost its battle against BIPA's steep consent requirements. After unsuccessfully arguing that the First Amendment protects the company's right to collect "publicly available information," Clearview AI accepted terms that it cannot sell its databases to private individuals or businesses in the US or to Illinois law enforcement agencies.[17] The company plans to pivot. "There are a number of other consent-based uses for Clearview's technology that the company has the ability to market more broadly," said the company's CEO.[18]

Consent has forced surveillance capitalism companies back to the drawing board, but privacy scholars and advocates were already at the drawing board. A new era of technology governance is underway.

CONSENT AS A BACKSTOP: THE NEW DATA SUBJECT

Wins for consent have occurred at a time when consent has fallen out of vogue in the privacy community.[19] Privacy scholars were and are exhausted from explaining and trying to solve insurmountable privacy self-management problems—problems I have reframed as failures derived from conflating computer characters and political capture. Legal scholar Ari Waldman articulates the current position of many privacy scholars most effectively: "The performative nature of individual rights in privacy law, which has habituated us into thinking managing our privacy is an individual's responsibility, has also allowed industry to weaponize our exercise of those rights to undermine our privacy."[20] Waldman writes, "To the extent that second wave privacy laws offer individuals additional rights to access, correct, delete, and port information, they sit within a long tradition of privacy laws focused on atomistic personal autonomy and choice. Most scholars agree that this conception of privacy is outdated and incompatible with today's information ecosystem."[21] Consent has become a dirty word for those inside the privacy community and increasingly for those outside. The message is that consent doesn't work—consent got us into this mess and it's not going to get us out.

On Twitter, UK privacy expert Michael Veale responded to the rejection of consent, explaining that if you want to get rid of consent because it will always be exploited, "You need a scalable and rigorous alternative to companies themselves undertaking an internal balancing test to decide on data use in their favour . . . [R]eaching into the organisation is tricky. Workable alternatives that aren't less protective are few."[22] Of those few alternatives, the two that have gotten the most traction are risk assessments and fiduciary responsibility.

The legacy of the vulnerable data subject lives on in European data protection law, where efforts are underway both to establish meaningful consent and to legitimize scalable, rigorous risk assessments. Under article 35 of the GDPR, data controllers/processors must carry out data protection impact assessments (DPIA) for high-risk processing. DPIAs represent an evolution of privacy impact assessments already required in some domestic data protection law and in article 20 of the Data Protection Directive, which required data protection agencies to assess certain data processing when they were notified. DPIAs in the GDPR move the risk assessment process from the DPAs to the data controllers, who have to produce a report for "blacklisted" processing that involves information like biometric, genetic, location, and employee monitoring data.[23] In addition to risk assessments, privacy scholars Woodrow Hartzog and Neil Richards, along with others, propose establishing a system of trust that would create a fiduciary duty for data holders.[24] Such a duty would obligate data collectors to pursue the best interests of the data subject with respect to data processing, which would in turn create opportunities to restrict design, alter financial incentives, and address exploitation. Anti-monopoly luminary Lina Khan and legal scholar David Pozen expressed skepticism and concern that such a system "abandoned a more robust vision of public regulation" and was enticing to Mark Zuckerberg, who called the proposal "thoughtful" and "intuitive."[25] The role that individuals should have in a system of fiduciary trust and whether individuals should have the power to object to or override the conditions involved with duties of care are details still being worked out. Both internal risk assessments and fiduciary duties could be powerful, but they may not be sufficient and could be abused over time, political change, and technical shifts. Consent can certainly be abused but so can internal processes that remove the individual from acting on their own behalf.

Stateside, privacy scholars still face an opt-out default so rigidly stuck in law, culture, and technical design that fighting against consent theater seems like a waste of time. California, however, offers a potential example of how consent and defaults can still become meaningful even in an opt-out America. In 2018, California managed to pass consumer privacy legislation that it then transformed into data protection legislation, but the strategy reads like a trick play—wild and impossible to replicate. Wealthy Bay Area real estate developer, Alastair Mactaggart, rubbed elbows at a cocktail party with a Google engineer who told him he would be horrified by the amount of data collected on users. Unknown in the state or national political or tech scenes, Mactaggart had "literally a shower thought" to put data privacy up to a vote by the people of California.[26] With the help of another parent from his kid's school, but single-handedly bankrolling the effort, Mactaggart began working on a data privacy ballot initiative.[27] He acquired the help of Ashkan Soltani, well-known in tech politics as a savvy privacy expert. As the former chief technologist of the FTC and an active participant in the W3C's DNT efforts, Soltani knew how hard Silicon Valley companies would push policymakers to prevent meaningful change to their operations.[28] By November 2017, the team presented Californians with a vote to slowly end third-party tracking, as well as a private right of action (meaning consumers could sue companies directly for violating the law as opposed to filing a complaint with a state agency). Tech companies scrambled to pressure state legislators to instead pass a less stringent law, which could be changed over an amendment period. Lawmakers hastily passed the California Consumer Privacy Act (CCPA) of 2018. The CCPA went into effect in 2020 but was amended that year by another ballot initiative, also led by Mactaggart, called the California Privacy Rights Act (CPRA). Both laws incorporated opt-out consent models, but CPRA moved the statute from a consumer privacy law to one that looked more like European data protection law (despite the fact that it still refers to consumers).

The CPRA amendments added the creation of a California Privacy Protection Agency; rights to correct, object, restrict, and reject data processing; and a definition of consent similar to the GDPR. Both CCPA and CPRA required businesses that handle personal data in certain ways to include a prominent "Do Not Sell My Personal Information" link on their home pages.[29] More importantly, the California attorney general specified that consumers may communicate their preference to opt out through an automated signal.[30] In other words, while companies have individual opt-out buttons on their sites, Californians should also be able to automatically send a signal to all sites or to a group of selected sites by changing their browser settings. To do this, a group spearheaded by Soltani is developing Global Privacy Control (GPC) as a proposed specification. The GPC can currently be sent by browser extensions and is respected by a small number of partner sites like the *New York Times* and the *Washington Post*, and the group submitted the standard to the W3C in hopes that it would be built into all major browsers. Many correctly call this another go at DNT, but the California attorney general didn't want to use the old DNT mechanism because "the majority of businesses disclose that they do not comply with those signals . . . [and] that businesses will very likely similarly ignore or reject a global privacy control if the regulation permits discretionary compliance."[31] Sites that want to ignore the signal may do so, but only if the consumer can "revoke consent as easily as it is affirmatively provided," which sounds like European efforts.[32] And in wording that is reminiscent of the DNT saga, according to the CPRA the signal "must represent a consumer's intent and be free of defaults constraining or presupposing that intent," meaning California consumers have to choose.[33] If and when GPC is integrated into browsers, the consumer choice can be legally enforced by California's very own data protection agency, as can the consent interfaces used to work around the selected setting. GPC could essentially be a DNT setting with Californian legal teeth, overcome only by European-style consent.

Consent for data subjects won some powerful victories at just the right moment to remain relevant in technology policy debates. These victories have driven policymakers and tech companies to the table to consider new governance structures. The trick is designing governance structures that can address the social harms of surveillance capitalism companies, while also accounting for potentially powerful and necessary individual rights. Data subject consent still must be bolstered with other protections. The Illinois BIPA law refers to "subjects" and as a kind of data protection law, it does more than require initial and meaningful consent to exercise choice. The law flat out bans the selling of or profiting from biometric information and allows individuals to bring actions directly against companies in court when violations occur. GDPR and CPRA provide higher standards for consent and complementary rights of access, rectification, and deletion, as well as rights to object and withdraw consent, but they also create considerable obligations, including risk assessments, data minimization requirements, design specifications, and record keeping for processing and contracts. This arrangement of individual rights complemented with obligations on the data collector and processor form what legal scholar Margot Kaminski calls binary governance.[34] The new data subject remains a character of control but also one protected by stricter obligations and risk assessments, thereby strengthening the data subject through this binary governance approach.

CONSENT AS PRIVACY CIVICS: THE NEW USER

No company has done more to breathe life back into digital consent than Apple. In June 2020, Apple announced it would change the data sharing default on the iPhone through operating system update number 14.5. On older versions of the iPhone, Apple allowed advertisers to track users with a unique ID For Advertisers (IDFA) string that effectively worked as a cookie but had a much longer life span and had to be manually changed by users (the Android system is similar, using a device identifier called Google Advertising ID (GAID)). The new opt-in system is part of the App Tracking Transparency (ATT) program, which prompts users to select between two options in a pop-up that asks "Allow 'App' to track activity across other companies' apps and websites?" In order to use an app, users must select either "Allow" or "Ask App Not to Track" the first time they open it (see figure 6.1).

Facebook lashed out over the change. "We're standing up to Apple for small businesses everywhere," read the full-page ad Facebook took out in the December 16, 2020 editions of the *New York Times*, the *Wall Street Journal*, and the *Washington Post*.[35] The ad continued, "Apple's forced software update [would] limit businesses' ability to run personalized ads and reach their customers effectively." The update would, according to the ad, decrease sales by 60 percent for every dollar spent. The Dutch version declared, "Nieuwe klanten werf je via Facebook"—"you recruit new customers via Facebook." The next day ads ran with the headline "Apple vs. the free internet," claiming that Apple's update would "change the internet as we know it—for the worse."[36] According to Facebook, Apple's forced update would mean content providers (like "cooking sites and sports blogs") would have to start charging subscriptions fees and in-app purchases, reducing the quality of free content. Facebook's ads concluded, "We hear your concerns, and we stand with you."

Facebook tried to appeal to people by emphasizing social enterprises like small business and free information services, but instead of pushing individuals to consent to data sharing that would support those social enterprises, it encouraged them to fight against a change to the default setting. Defaults can be understood to promote a particular social outcome. Opt-out defaults are used to increase organ donation rates, for instance, because people don't change the defaults.[37] Choice architectures can nudge people to decisions to meet their goals and build good habits, but default settings

FIGURE 6.1
Tracking consent interface on iPhone after iOS 14.5 update.

reveal the values of the system. Despite hearing consistent and growing concern from the public about the abuse of personal data, US policymakers and Silicon Valley maintained an opt-out default that allowed unfettered information collection and sharing, pitched as an arrangement for free commercial internet services.

The opt-out default did more than serve the social enterprise of the free market. Policymakers and political campaigns had gotten into the tracking game too. Although politicians have been strategizing with the help of computer data since even before the secret "People Machine" (an overhyped FORTRAN program that processed voter data) used to orchestrate a win for John F. Kennedy over Richard Nixon in 1960, President Barack Obama's 2008 campaign brought the practice into the "big data" era.[38] Obama famously used data to target voters and folded these tactics into his branding as a young, innovative, and digitally savvy candidate. In 2009, NationBuilder launched to make this style of campaign more readily available to all candidates. NationBuilder quickly became *the* political software and platform used by political campaigns across the country and then around the world. The platform is notable because it can easily incorporate third-party data from commercial data brokers and social sites like Twitter and Facebook.[39] Everyone uses the platform—from Donald Trump and Brexit advocates to Emmanuel Macron and the NAACP. Joe Green, a cofounder and the president of NationBuilder in its first few years, was Mark Zuckerberg's roommate at Harvard and worked with him on early iterations of the social media site.[40] Green is considered a social entrepreneur, having started the Facebook app Causes with Napster founder and Facebook backer Sean Parker to encourage activism and philanthropy, and is referred to as Zuckerberg's "political wingman," having led Facebook's advocacy group on immigration reform in Washington.[41] NationBuilder distances itself from shady political data services that are "in the business of harvesting your data without your consent," and affirms that it is "in the business of allowing organizations to manage the data that people have opted to give them."[42] Or, more accurately, not opted out of giving them.

But consent on the iPhone is an example of individual users actively rejecting the status quo established through years of an opt-out default. The iPhone update has produced a powerful message against Facebook's data collection practices. Under the new ATT policy, 96 percent of iPhone users "Ask App Not to Track," meaning only 4 percent opt to allow apps to track their activity across other companies' apps and websites.[43] When presented with the opportunity, iPhone users reject the collective enterprise of free apps, sites, and services based on invasive, opaque, and dodgy data practices. Facebook shares dropped and continued to slide after news that the company was struggling to contain the damage from Apple's iOS update and would continue to struggle under the system for years.[44] Even as many technology companies thrived during the pandemic, Apple's ATT system cost Facebook $10 billion in revenue and forced significant change in the company.[45] In fall 2021, Facebook rebranded itself as Meta to pursue the metaverse, an ambitious vision of a virtual reality network.[46] In the metaverse, Facebook would no longer sit precariously atop Google and Apple's hardware (e.g., mobile phones and computers) or software (e.g., operating systems and browsers), but would operate at lower layers in a network infrastructure, isolated from Tim Cook's political whims.

In an interview with Kara Swisher, Cook said Apple has added privacy features every year and that "if you were designing such a system from scratch today, of course it should be your decision for what happens to your data."[47] But the timing of Apple's iOS change is no coincidence. Cook and Zuckerberg have feuded publicly over this issue for the last few years, and Cook was particularly critical after the Cambridge Analytica scandal, saying on primetime television:

> We're not going to traffic in your personal life. I think it's an invasion of privacy . . . [P]rivacy to us is a human right. It's a civil liberty, and in something that is unique to America, you know, this is like freedom of speech and freedom of the press and privacy is right up there for us . . . Everybody should know what they're giving up. And not only the specific data points, but . . . when I know this, plus this, plus this, plus this I can infer a whole bunch of other things and that can be abused. And it can be abused against our democracy.[48]

When asked what he would do in Zuckerberg's position, Cook responded, "What would I do? I wouldn't be in this situation."[49] On the Ezra Klein Show podcast, Zuckerberg responded to Cook's criticism, "I think it's important that we don't all get Stockholm syndrome and let the companies that work hard to charge you more convince you that they actually care more about you. Because that sounds ridiculous to me."[50]

Cook's response about timing also failed to mention pressure from regulators and the public, who have expressed frustration with Cook's fellow tech CEOs. The March 2018 Cambridge Analytica scandal (that brought Zuckerberg to Washington in the introductory chapter of this book) shook users as voters after the 2016 election, and people wanted answers. Funded largely by right-wing donor Robert Mercer, Cambridge Analytica built voter profiles using data scraped from Facebook. Whistleblower and director of research at Cambridge Analytica, Christopher Wylie, explained that Facebook users took surveys that captured their data and, if they had not adjusted their privacy settings, data from users' friends and contacts: "[B]y filling out my survey, I capture 300 records on average . . . And so that means that, all of a sudden, I only need to engage 50,000, 70,000, 100,000 people to get a really big data set really quickly, and it's scaled really quickly.

We were able to get upwards of 50 million plus Facebook records in the span of a couple of months."[51] Wylie alleged that President Trump's campaign executive and right-wing media leader Steve Bannon, who sat on the board of Cambridge Analytica, coveted the harvested data to wage "his culture war."[52] User data was also directly and indirectly tied to Russian interference with the US elections and other campaigns around the world. Directly, Russian troll farms like the Internet Research Agency targeted users with polarizing propaganda by using the same data-targeting techniques on platforms used by domestic politicians.[53] Indirectly, some well-connected purveyors of US political data and information services allegedly used Russian storage, email, and search companies, potentially exposing the election information to access by Russian intelligence.[54]

The Democrats were already looking into Russian interference in the 2016 election and demanded Zuckerberg testify before Congress. When Zuckerberg came to answer questions only a month after initial reporting on Cambridge Analytica, policymakers had gathered not to talk about innovation or free internet or decentralized networks but to make a show of Facebook's abuse of the way personal information connects to democratic processes. Delaware Senator Christopher Coons pressed the problem succinctly to Zuckerberg:

> Facebook claims that advertising makes it easy to find the right people, capture their attention and get results and you recognize that an ad-supported service is, as you said earlier today, best aligned with your mission and values. But the reality is, there's a lot of examples where ad targeting has led to results that I think we would all disagree with or dislike or would concern us. You've already admitted that Facebook's own ad tools allow Russians to target users, voters based on racist or anti-Muslim or anti-immigrant views, and that that may have played a significant role in the election here in United States.[55]

Since the 2016 election, public discourse has continued to evince concerns that personal information is vulnerable to political manipulation. In April 2021, three years after Zuckerberg testified, a Senate judiciary subcommittee heard testimony about how data-driven innovations have been "harnessed into algorithms designed to attract our time and attention on social media and the results can be harmful to our kids' attention spans, to the quality of public discourse, to our public health, and even to our democracy itself."[56] Suddenly, cookies weren't just about ads for shoes following users around the internet. They were used to manipulate our families and friends—weapons used to attack democracy.[57] Apple implemented its iPhone privacy update after publicly promoting the role that privacy plays in supporting democracy. The iPhone update is perhaps proof of concept that, under the right conditions, user consent can serve privacy and privacy can serve democracy.

As part of larger scholarly efforts to articulate and understand the collective nature of privacy, Salomé Viljoen highlights what she calls "data relations." Viljoen argues that existing critiques in data debates miss the collective operations of datafication and injustices that exceed individual control or economic fairness. Viljoen showcases these shortcomings by asking what's wrong with the US military purchasing location data from Muslim Pro, a Muslim prayer app that uses GPS to provide locally accurate times for the five daily prayers. Neither individual control nor economic unfairness account for the "significance of the data flow between Muslim Pro and the US military . . . [that] drafts users—faithful Muslims—into the project of their fellow Muslims' oppression."[58] Viljoen asks, "what projects are worth being drafted into?" A follow up question might be, "And how do we express what projects are worth [users] being drafted into?" Although Tim Cook's portrayal of the iPhone's consent system sounded like early 2000s retro privacy discourse, his consent interface is one way to express rejection of projects we do not want to be drafted into, which is particularly important when defaults serve those in power and are therefore hard to flip.

User consent in a world that recognizes the collective value of privacy could be wielded differently than consent in a privacy era defined by individual self-interest. The user remains tied to civic concepts like privacy, anonymity, speech, and the potential of networks. Consent to authorize civic ends could allow users to fight for interface design akin to how voting rights advocates battle for ballot design to promote fair elections. Users have shown their unwillingness to authorize certain collective enterprises like biometric tracking and surveillance capitalism. For years, Silicon Valley and captured politicians have promoted consent as a governance tool, but as implemented on the iPhone, such consent enthusiasm may disappear.

CONSENT AS HEALTHY COMPETITION: THE NEW PRIVACY CONSUMER

Consumer protection is a restricted path, but it may be the avenue that can produce the most immediate results and should certainly be bolstered, particularly considering recent scrutiny of tech companies' monopoly powers. Following its initial announcement about the retirement of cookies, Google published a second post that pushed the date back to the end of 2023.[59] The delay helped make it clear that Google would be working with the W3C community to create new standards, not redesigning the system to serve one company. Google proposed a "Privacy Sandbox" to replace third-party cookies. The Privacy Sandbox is a collection of proposals meant to get at different tasks previously taken up by third-party cookies.[60] Under international antitrust scrutiny, Google submitted the collection to the W3C's Improving Web Advertising Business Group, established in 2017 to embrace a "'Web advertising by design' approach that includes the needs of all parties involved, that acknowledges the reality today of the advertising-funded Web sites, and that places privacy, security, accessibility, and user experience all in the forefront."[61] Putting the Sandbox in the W3C allows Google to act in collaboration with its competitors. The timeline for consensus is slower.[62] It gives related industries and interested parties more time to hammer out the details of a post-cookie internet. Nonetheless, Google's behavior within the W3C drew the attention of the UK's Competition and Markets Authority, which was concerned that changes would "undermine the ability of publishers to generate revenue and undermine competition in digital advertising, entrenching Google's market power."[63]

Calling the whole thing a Pandora's Box, James Rosewell, the CEO of data analytics company 51Degrees, asked, "Should web browsers really become implementation mechanisms of specific government regulation?"[64] Like the IAB in DNT debates of the past, Rosewell attempted to keep big companies and the law separate from technical standards. He and others find working within the W3C less than productive. Rosewell stated, "I've been—I don't know what the right word is—somewhere between upset and shocked at just how much of a sort of vigilante group the W3C truly is."[65] As another ad-tech executive put it, "When you're going up against powerful companies that are very entrenched in the W3C, and you're saying something they don't want said, it can feel as though you're being gaslit, given contradictory information on rules that aren't applied later."[66]

Others criticized Google's participation in the W3C too. Google was the only member of W3C's Privacy Interest Group (PING) to vote "no" to an update to its charter in 2019.[67] The change, which required a unanimous favorable vote, would have expanded the power of PING to block new specifications that threaten web privacy. Narayanan and Mayer accused Google of "privacy gaslighting" for arguing that ending third-party cookies could be bad for privacy and for claiming that the company was trailblazing privacy-preserving advertising:

> Apple and Mozilla have tracking protection enabled, by default, today. And Apple is already testing privacy-preserving ad measurement. Meanwhile, Google is talking about a multi-year process for a watered-down form of privacy protection. And even that is uncertain—advertising platforms dragged out the Do Not Track standardization process for over six years, without any meaningful output. If history is any indication, launching a standards process is an effective way for Google to appear to be doing something on web privacy, but without actually delivering.[68]

Google has defenders, though. Ben Savage, a software engineer at Facebook who actively participates in W3C efforts, accused Apple of "egregious behavior" for changing tracking policies with a quick iOS update and without consultation. Savage suggested Apple wanted to drive apps on the iPhone away from ad-based models and toward in-app purchases (of which Apple takes a cut). He argued that Facebook operates "in the open, announcing their intended privacy goals, proposing new APIs publicly, soliciting input, publicly responding to feedback, allowing people to test," whereas Apple "decided to just blow up everything first . . . and only start thinking about what to replace it with later."[69] Simply ending third-party tracking works for the Facebooks, Googles, and Apples of the world, Savage and Rosewell agree, but what about other apps and sites?[70] There will be a new set of winners and losers. Ad network executives pushed back hard when Google announced the end of third-party cookies. "[Google] has been successful by being able to take in as much data as possible and not allowing it to leave," said a representative of Digital Content Next. "The walled gardens will have higher walls."[71]

The capacity of a few big platforms to hook users is a major policy concern because it risks not just civic destruction but also consumer manipulation. As scrutiny of potential monopolies increases so too does scrutiny of consumer choice. Documentaries like the Netflix film *The Social Dilemma* (2020) and a *60 Minutes* segment on children and screentime were incredibly popular in recent years, appealing to audiences desperate to unplug and eager to know why doing so is so hard. In summer 2020, Congress questioned the CEOS of Google, Amazon, Apple, and Facebook—all companies criticized for the path they took to gaining their dominant positions—about whether they manipulated consumers from that position of dominance. Policymakers interrogated Amazon about using data from third-party sellers to launch their own private label products, Facebook about threatening to clone a company while attempting to acquire it, and Google about whether the company promoted its own sites and services above others.[72] Google CEO Sundar Pichai explained what had been repeated throughout the five-hour session, "Just like other businesses, we try to understand trends from data which we can see and we use it to improve our product for our users."[73] A year later, Google and Facebook executives were brought back to Washington for a hearing on competition and privacy. Republican Senator Marsha Blackburn probed how data was used to manipulate consumer experience with content and accused Facebook of abusing data collected on teenagers while turning a blind eye to claims of addiction and damage to their mental health.[74]

In the hands of a consumer protection prosecutor, manipulation might be considered an unfair practice. Asserting affirmative, prior consent for the US privacy consumer would have seemed wildly unrealistic only a few years ago given the American privacy consumer of the early web, but a bipartisan techlash opened the way for legal innovation. The FTC has recently pursued a new set of violations referred to as "dark patterns." Dark patterns are "user interface designs that manipulate consumers into taking unintended actions that may not be in their interest."[75] Insofar as they don't make it obvious that consumers can change preferences or that they require a consumer to click through numerous confusing screens to learn about or reject data practices, cookie pop-ups and other privacy settings certainly fit that definition. Indeed, numerous participants in the FTC's Bringing Dark Patterns to Light workshop repeatedly pointed out the problems with cookie consent mechanisms.[76] Given the vast amount of research that privacy scholars have produced about default settings and privacy self-management, the FTC could consider defining dark patterns for consumer data collection using the French rule: it must be just as easy to consent as to refuse.

In addition to recognizing dark patterns, the FTC recently recognized a new kind of unfairness harm: circumventing browser privacy settings. In November 2020, the FTC went after Zoom for bypassing browser safeguards in violation of section 5 of the FTC Act. Privacy law scholars Solove and

Hartzog explained how the consent decree represented both a disappointing rite of passage for Zoom and new developments at the FTC.[77] Most of the decree details the broken promises and false assertions Zoom made about its security practices—the FTC has been in that game for a long time and Zoom had to stop making these false claims—but then the decree turned to Zoom's circumvention of the Safari prompt that made sure consumers chose to enter a video conference. The complaint explained that without the prompt, "[O]ne wrong click could expose consumers to remote video surveillance by strangers through their computers' webcams."[78] The new privacy consumer needs more than what privacy scholar Chris Hoofnagle calls "the FTC's catch and release program," and commissioners Rebecca Slaughter and Rohit Chopra plan to provide it.[79] Their dissents in the Zoom settlement make clear they do not think the FTC went far enough. They want deeper, more effective, and more impactful FTC prosecutions that aren't confined to a narrow notion of security.[80] Under a democratic president, Chopra moved on to direct the Consumer Financial Protection Bureau, and Slaughter has been joined at the FTC by antitrust reformer Lina Khan and social justice privacy advocate Alvaro Bedoya.[81] The catch and release program may follow the cookie into retirement.

NEW CHARACTERS FOR TECH GOVERNANCE

Even the best versions of consent for any computer character provide limited governance over technology companies or data-heavy organizations, something noted by those who want to divest from consent. We will need new computer characters to fight for (and as) in order to achieve broader governance goals. I suggest two computer characters to promote a more expansive strategy: the non-user and the members of the public.

The non-user is a character who doesn't want to opt out of the system; they want or need to avoid the system.[82] Instead of focusing on consent and emphasizing control, authorization, or choice, recognizing non-users is an opportunity to address a range of people who reject or who are rejected by a technical system. In 2003, STS scholar Sally Wyatt articulated the need to account for non-users who are not potential users or not-yet users, the only alternatives to the user provided in much of the discussion of the digital divide.[83] Wyatt sketched out a set of four actors with multiple motivations that have led them to be non-users: resisters, who have never been online because they do not want to be; rejecters, who have left voluntarily; excluded, who are not able to get access; and expelled, who have lost access.

The non-user accounts for both those who can use a thing but don't want to, and those who can't use a thing because they were left out of the design or prohibited from the system. Building in alternatives for those who do not support a company, system, or initiative forces friction into the design process that provides beneficial, if costly, foresight. Georgetown University's Ethics Lab explains how a "user-centered" design framework trains technologists to see non-users as adversarial or convertible and does not allow for recognition of considerable moral concerns. Designing for non-use provides an offramp for poor or damaging ideas. In promoting a non-user-centered approach, the Ethics Lab supports, for example, aversion to biometric password systems on smartphones as essential feedback, rather than a problem to be solved through interface design and encryption.[84] As designer Christine Satchell and computer and social scientist Paul Dourish eloquently explain, "[N]on-use is not an absence or a gap; it is not negative space. Non-use is, often, active, meaningful, motivated, considered, structured, specific, nuanced, directed, and productive."[85] Those who want to maintain cash payments, dumb phones, and iPad-less childhoods may find the non-user to be a powerful character to play and the fight for non-use to be a fruitful campaign.

Non-use may also help account for barriers to use in disability communities. Technology law scholar Blake Reid challenges the place-based metaphors (i.e., places of public accommodation like restaurants, theaters, shops, hotels, banks, zoos, libraries, etc.) employed to make websites accessible, while leaving large chunks of the internet inaccessible.[86] Reid suggests moving away from the perspective of user experience and instead assessing accessibility across the layered internet architecture, identifying larger challenges hidden in the stack. Rejecting universal design, inclusive designers like Liz Jackson call on us to "honor the friction of disability." Jackson explains that disability design has fostered touch screens, bicycles, email, audiobooks, and captioning. She wants disabled people to be designers, not just users and testers, and for design training to engage directly with the creative culture of the disabled community.[87] In a closely related network, design justice advocates have laid out principles that embrace the disability justice movement's slogan "nothing about us without us."[88]

A member of the public is another computer character who may support societal goals. Members of the public engage in activities that important social institutions rely upon, and thus are given certain roles, such as serving on committees, and powers, such as requesting information from the government using freedom of information processes, known as the Freedom of Information Act (FOIA) in the US. FOIA is a federal law that allows the public to access records held by federal agencies. Although both FOIA and the Privacy Act allow people to request records from the government, FOIA came into being to empower effective watchdogs, whereas the Privacy Act passed to correct records and control personal information in an essential system. FOIA passed after a decade of relentless work by California Congressman John Moss (a "kind of Mr. Smith gone to Washington") and eight years before the more hastily promulgated Privacy Act of 1974.[89] I argued in chapter 2 that the Privacy Act and other first-generation data protection laws were a product of their very particular moments in technical, cultural, and political histories, and FOIA is similarly a product of its time. Despite repeated references to Founding Fathers, journalism scholar Michael Schudson points out, "[N]othing remotely like [FOIA] appeared in the early American republic."[90] Schudson describes how Moss's dogged efforts were bolstered by a growing freedom of information movement led by journalists fighting government secrecy at the state level in the 1960s. Both Congress and the media were concerned about the power of a growing federal bureaucracy and developed a "right to know" to keep it in check.

In what Schudson refers to as FOIA's poetry, the act in fact allows "any person" (not members of the public) to make a request. While it relieves the FOIA officer from verifying the identity of the requester and thus frees them to assess the request on its merits, any person is also "a richer, more universal, and more complete rendering of democratic equality than the founders ever endorsed."[91] Any person may not capture the social values of today's privacy advocates but should at least serve as inspiration. Any person certainly struggles to make FOIA work for them. Schudson details the many and ongoing critiques of FOIA, while legal scholar Margaret Kwoka uncovers some clear problems in her empirical study of FOIA requests.[92] Federal agencies receive almost a million FOIA requests a year from a wide range of actors, including reporters, concerned citizens, nonprofits, and community organizations seeking information to hold the government accountable. Kwoka highlights non-oversight requesters—those who are seeking information not for public oversight but for private interest—as a major source of FOIA problems. Many of these non-oversight requests are trying to get information about corporate competitors and industry activity; others request information about their own records. Kwoka argues for structural changes to FOIA, including a specialized oversight body that can serve as an intermediary between agencies and the public. Despite FOIA's flaws, the law was globally influential, spurring an "openness revolution" that has started to reach multinational corporations.[93] Fighting as members of the public or as any person for access to information about data controllers and processors, platform operators, and dominant technology companies would represent an effort at governance as opposed to individual control. Learning from efforts to reform FOIA as it is applied to the government, any person or members of the public could fight for a FOIA applied to Big Tech.

These potential computer characters would be intentionally, politically made-up people, conjured into being not by sophisticated politicians or advertising experts but to serve public ends and social goals. They might help us transition to a more social practice of privacy, reinforced by technology policy and design. The user, data subject, and privacy consumer are worth rehabilitating as well. Although these now classic characters are entrenched, their histories, if untangled, can make them powerful tools for supporting societal goals.

CONCLUSION

So that's the long story of cookies, which is really a history of transatlantic imaginaries across iterations of technical change and the role of individual consent within those politics. Existing accounts describe a computational past where consent worked to protect people and keep technology accountable, invoking a wave of increased information overflow as a force that finally broke the previously effective use of consent. Neoliberal critiques are part of that largely inaccurate and unactionable story, but not the whole story. The story of cookies—and digital consent—is much richer and provides alternative interventions for reimagining sociotechnical futures.

By focusing on the computer characters at the heart of cookie law, digital consent can be seen as layered, local, and changing. Each derives from a distinct area of law—data protection, communication privacy, and consumer protection—that engaged in its own legal construction of people and computer technology. The data subject was constructed as a compromised, vulnerable character that required giving up data for essential services and who was given powers to control their data to avoid being controlled by those employing the power of computers. The data subject was created for computers, and the user for networks. The user as a computing character was tied to histories of citizenship and speech in communication networks that were transforming into digital networks. Far less passive than the data subject, the user was sophisticated enough to authorize access to a system. If they were too sophisticated, working at the margins of authorization, users became hackers. All-American, the privacy consumer was made up to promote consent as choice for the commercial web as it developed. Following their journeys serve as crucial means to untangle the complex political failing of cookie policy and our understanding of digital consent. It may also inspire the creation of new computer characters.

Wielded at a particular historical moment, consent drove cookies into retirement and in doing so drove policymakers and tech companies to reconsider socially desirable business models that support a contested set of values. Privacy scholars don't have all the answers, but they have worked to prioritize values and provide clarity on appropriate technical design, integration, and use. These efforts should not discard character development when trying to fight for a worthy sociotechnical reimaginary.

NOTES

SERIES EDITOR'S INTRODUCTION

1. D. Mills, "Distributed-Protocol Authentication Scheme," RFC 1004, April 1987, https://www.rfc-editor.org/info/rfc1004.

2. O. Sury, W. Toorop, D. Eastlake III, and M. Andrews, "Interoperable Domain Name System (DNS) Server Cookies," RFC 9018, April 2021, https://www.rfc-editor.org/info/rfc9018.

CHAPTER 1

1. "The Weekly Dish," *Washington Post,* March 7, 2001, https://www.washingtonpost.com/archive/lifestyle/food/2001/03/07/on-the-fridge/0f79245a-a0ed-4336-a558-e55485359436/.

2. Nora A. Draper, *The Identity Trade: Selling Privacy and Reputation Online* (New York: NYU Press, 2019), 68–73.

3. John Schwartz, "As Big PC Brother Watches, Users Encounter Frustration," *New York Times,* September 5, 2001, sec. A, 1.

4. Matt Burgess, "How to Avoid Those Infuriating Cookie Pop-Ups," *Wired,* May 22, 2021, https://www.wired.com/story/avoid-cookie-popups-gdpr/.

5. See chapter 4 for a full discussion of the origin story of the term "cookie."

6. Julia Angwin, "The Web's New Gold Mine: Your Secrets," *Wall Street Journal,* July 30, 2010, https://www.wsj.com/articles/SB10001424052748703940904575395073512989404.

7. Aaron Cahn, Scott Alfeld, Paul Barford, and S. Muthukrishnan, "An Empirical Study of Web Cookies," *Proceedings of the 25th International Conference on World Wide Web* (April 2016): 891–901, https://doi.org/10.1145/2872427.2882991.

8. Cahn et al., 891–901.

9. Paul Calvano, "An Analysis of Cookie Sizes on the Web," PaulCalvano.com, July 13, 2020, https://paulcalvano.com/2020-07-13-an-analysis-of-cookie-sizes-on-the-web/. The HTTP Archive is an open-source, community-supported project that stores details for every HTTP request and response for approximately 5.8 million home pages. Calvano is a co-maintainer of the archive.

10. Cory Doctorow, "Consent Theater," *OneZero*, May 20, 2021, https://onezero.medium.com/consent-theater-a32b98cd8d96.

11. *Data Privacy and Protection: Hearings before the Senate Committee on Commerce, Science, and Transportation and the Senate Committee on the Judiciary*, 115th Congress (April 10, 2018) (testimony of Mark Zuckerberg, CEO of Facebook), video available from C-SPAN: https://www.c-span.org/video/?443543-1/facebook-ceo-mark-zuckerberg-testifies-data-protection.

12. Daniel J. Solove, "Privacy Self-Management and the Consent Dilemma," *Harvard Law Review* 126, no. 7 (2012): 1880–1884.

13. Aleecia M. McDonald and Lorrie Faith Cranor, "The Cost of Reading Privacy Policies," *I/S: A Journal of Law and Policy for the Information Society* 4, no. 3 (2008): 543–568.

14. Helen Nissenbaum, professor of information science at Cornell Tech, refers to this as the "transparency paradox." Helen Nissenbaum, "A Contextual Approach to Privacy Online," *Daedalus* 140, no. 4 (2011): 32–48, https://doi.org/10.1162/DAED_a_00113; Corey A. Ciocchetti, "The Future of Privacy Policies: A Privacy Nutrition Label Filled with Fair Information Practices," *John Marshall Journal of Computer & Information Law* 26, no. 1 (2008): 1–32; Patrick Gage Kelley, Joanna Bresee, Lorrie Faith Cranor, and Robert W. Reeder, "A 'Nutrition Label' for Privacy," *Proceedings of the 5th Symposium on Usable Privacy and Security* (July 2009): 1–12, https://doi.org/10.1145/1572532.1572538; Marcus Moretti and Michael Naughton, "Why Privacy Policies Are So Inscrutable," *The Atlantic*, September 5, 2014, https://www.theatlantic.com/technology/archive/2014/09/why-privacy-policies-are-so-inscrutable/379615/.

15. Margaret Jane Radin, *Boilerplate: The Fine Print, Vanishing Rights, and the Rule of Law* (Princeton, NJ: Princeton University Press, 2012); Frederik Zuiderveen Borgesius, "Informed Consent: We Can Do Better to Defend Privacy," *IEEE Security & Privacy* 13, no. 2 (2015): 103–107, https://doi.org/10.1109/MSP.2015.34; Elizabeth Edenberg and Meg Leta Jones, "Analyzing the Legal Roots and Moral Core of Digital Consent," *New Media & Society* 21, no. 8 (2019): 1804–1823, https://doi.org/10.1177/1461444819831321.

16. Kashmir Hill, "I Cut the 'Big Five' Tech Giants from My Life: It Was Hell," *Gizmodo*, February 7, 2019, https://gizmodo.com/i-cut-the-big-five-tech-giants-from-my-life-it-was-hel-1831304194.

17. Solove, "Privacy Self-Management and the Consent Dilemma."

18. Woodrow Hartzog and Neil Richards explain, "We live in a society that lionizes individual choice in the many social roles we play every day, whether as consumers, citizens, family members, voters, lovers, or employees. Consent reinforces fundamental cultural notions of autonomy and choice. It transforms the moral landscape between people and makes the otherwise impossible possible." Neil Richards and Woodrow Hartzog, "The Pathologies of Digital Consent," *Washington University Law Review* 96, no. 6 (2018): 1462. Philosopher David Johnston explains that, although "we remain within a long historical moment in which, in western societies, the notion that individuals should be subject only to those obligations to which they have freely given their consent retains enormous power," the maintenance of order has not always relied on individual consensual relationships. In the earliest Greek philosophical writings and Roman laws, individual consent played a large role, though very few individuals could consent given their social status. In the fifth century, confidence in the individual diminished dramatically. Kinship ties became the predominant source of obligations and entitlements. During this period, standardization set the price of goods, and the ordeal settled legal disputes. The rediscovery of Greek philosophy, an interest in Roman law, and the universality of Christianity combined to launch a revival of faith in individual consent during the early modern period. Immanuel Kant's ideas over the course of the 1700s proved most powerful in German law and culture. Simply put, Kant argued that freedom is the highest attainment in human affairs—an individual is not just inherently free but can attain freedom—and the decisions and agreements of fully autonomous individuals produce the best society. David Johnston, "A History of Consent in Western Thought," in *The Ethics of Consent: Theory and Practice*, ed. Franklin Miller and Alan Wertheimer (New York: Oxford University Press, 2010), 25–54. In Germany, the defense of consent is traced to the legal principle of *volenti non fit iniuria* (no wrong is done to one who consents). The German Basic Law (*Grundgesetz*) constitutionally enshrines the right to autonomy and self-determination. It is integral to German law that one is free to dispose of their legal interests, waive protections, and consent to harmful acts. By contrast, the French do not hold consent to neutralize liability, at least not for criminal liability. Kai Ambos and Stefanie Bock, "Germany," in *Consent: Domestic and Comparative Perspectives*, ed. Alan Reed and Michael Bohlander (Abingdon, UK: Routledge, 2017), 262–279.

19. John Kleinig, "The Nature of Consent," *The Ethics of Consent: Theory and Practice*, ed. Franklin Miller and Alan Wertheimer (New York: Oxford University Press, 2010), 3–24.

20. Heidi M. Hurd, "The Moral Magic of Consent," *Legal Theory* 2, no. 2 (June 1996): 121–124, https://doi.org/10.1017/S1352325200000434.

21. James Glave, "PGP Lets You Take Charge of Your Cookies," *Wired*, December 10, 1996, https://www.wired.com/1996/12/pgp-lets-you-take-charge-of-your-cookies/.

22. *Data Privacy and Protection: Hearings before the Senate Committee on Commerce, Science, and Transportation and the Senate Committee on the Judiciary*, 115th Congress (April 10–11, 2018) (testimony of Mark Zuckerberg, CEO of Facebook), video available from C-SPAN: https://www.c-span.org/video/?443543-1/facebook-ceo-mark-zuckerberg-testifies-data-protection; https://www.c-span.org/video/?443490-1/facebook-ceo-mark-zuckerberg-testifies-data-protection.

23. Ian Hacking, "Making Up People," in *Reconstructing Individualism: Autonomy, Individuality, and the Self in Western Thought*, ed. Thomas C. Heller, Morton Sosna, and David E. Wellbery (Redwood City, CA: Stanford University Press, 1986), 222–236; quote from Ian Hacking, "Kinds of People: Moving Targets," *Proceedings of the British Academy* 151 (2007): 285, https://www.thebritishacademy.ac.uk/documents/2043/pba151p285.pdf.

24. Ian Hacking, "The Making and Molding of Child Abuse," *Critical Inquiry* 17, no. 2 (1991): 253–288, https://doi.org/10.1086/448583.

25. Hacking, "Making Up People," 228.

26. Such work differs slightly from other work in the history of technology and privacy scholarship. For example, Dan Bouk writes of the "statistical individual" in his book *How Our Days Became Numbered*, wherein he details the evolution of the life insurance industry. Between the 1870s and 1930s, the industry's goal of writing as many life insurance policies as profitable pushed companies in a highly competitive environment to more accurately quantify mortality risk. To do so, companies expanded the measurement of factors that affect health. The expansion surfaced a statistical individual around which revolved tensions between individualized versus averaged risk and reading versus changing fate. Bouk's descriptive label differs slightly from normative labels. Dan Bouk, *How Our Days Became Numbered: Risk and the Rise of the Statistical Individual* (Chicago: University of Chicago Press, 2015). In her book *Configuring the Networked Self*, Julie Cohen criticizes the liberal self—autonomous, rational, and disembodied—as outdated and inaccurate. Cohen's "networked self" is "both irreducibly embodied and constituted partly by and through the technologies and artifacts that surround it." The networked self describes an aspiration that would lead to normative differences, as Cohen outlines the policies to support the flourishing of this self. Julie E. Cohen, *Configuring the Networked Self: Law, Code, and the Play of Everyday Practice* (New Haven, CT: Yale University Press, 2012).

27. E.g., the July 1973 *"Records, Computers and the Rights of Citizens," Report of the Secretary's Advisory Committee on Automated Personal Data Systems* (commonly known as the *HEW Report*) refers to the "data subject" in its Summary and Recommendations, stating, "If record-keeping systems and their data subjects were protected by strong safeguards, the danger of inappropriate record linkage would be small; until then there is a strong case to be made for discouraging linkage." Loi 78-17 du 6 janvier 1978 relative à l'informatique, aux fichiers et aux libertés (version consolidée

NOTES   195

au 27 août 2011) [Law 78-17 of January 6, 1978, on Information Technologies, Data Files and Civil Liberties (consolidated version as of August 27, 2011)], unofficial English translation at https://www.cnil.fr/sites/default/files/typo/document/Act78-17VA.pdf.

28. Directive 2002/58/EC of the European Parliament and of the Council of 12 July 2002 concerning the processing of personal data and the protection of privacy in the electronic communications sector (Directive on privacy and electronic communications).

29. California Consumer Privacy Act of 2018, Cal. Civ. Code § 1798.100 et seq. (2018).

30. Sheila Jasanoff and Sang-Hyun Kim, eds., *Dreamscapes of Modernity: Sociotechnical Imaginaries and the Fabrication of Power* (Chicago: University of Chicago Press, 2015).

31. Jack M. Balkin and Reva B. Siegel, "Principles, Practices, and Social Movements," *University of Pennsylvania Law Review* 154, no. 4 (2006): 927–950, https://doi.org/10.2307/40041288.

32. For an in-depth discussion of imagined regulatory scenes and technological change, see Margot E. Kaminski, "Technological 'Disruption' of the Law's Imagined Scene: Some Lessons from *Lex Informatica*," *Berkeley Technology Law Journal* 36, no. 3 (2022), https://btlj.org/wp-content/uploads/2023/01/0002-36-3-Kaminski_Web.pdf.

33. Lisa Messeri and Janet Vertesi, "The Greatest Missions Never Flown: Anticipatory Discourse and the 'Projectory' in Technological Communities," *Technology and Culture* 56, no. 1 (2015): 56, https://doi.org/10.1353/tech.2015.0023.

34. Federal Trade Commission, "Facebook Settles FTC Charges That It Deceived Consumers by Failing to Keep Privacy Promises," FTC press release, November 29, 2011, available at PR Newswire: https://www.prnewswire.com/news-releases/facebook-settles-ftc-charges-that-it-deceived-consumers-by-failing-to-keep-privacy-promises-134681358.html.

35. Chris Hoofnagle, "Facebook in the Spotlight: Dataism vs. Privacy," *Jurist*, April 20, 2018, https://www.jurist.org/commentary/2018/04/chris-hoofnagle-facebook-dataism.

36. Federal Trade Commission, "Statement of Chairman Joe Simons and Commissioners Noah Joshua Phillips and Christine S. Wilson, *In re Facebook, Inc.*," FTC public statements, July 24, 2019.

37. David Vincent, *Privacy: A Short History* (New York: John Wiley & Sons, 2016), 2.

38. Samuel D. Warren and Louis D. Brandeis, "The Right to Privacy," *Harvard Law Review* 4, no. 5 (1890): 193, 216, https://doi.org/10.2307/1321160.

39. William L. Prosser, "Privacy," *California Law Review* 48, no. 3 (August 1960): 383–423, 388–389, https://doi.org/10.15779/Z383J3C.

40. Daniel J. Solove, *Understanding Privacy* (Cambridge, MA: Harvard University Press, 2008).

41. Irwin Altman, "Privacy Regulation: Culturally Universal or Culturally Specific?," *Journal of Social Issues* 33, no. 3 (1977): 66–84, https://doi.org/10.1111/j.1540-4560.1977.tb01883.x.

42. Sarah E. Igo, *The Known Citizen: A History of Privacy in Modern America* (Cambridge, MA: Harvard University Press, 2018).

43. Michael Schudson, *The Rise of the Right to Know: Politics and the Culture of Transparency, 1945–1975* (Cambridge, MA: Harvard University Press, 2015); Amy Gajda, *Seek and Hide: The Tangled History of the Right to Privacy* (New York: Viking, 2022).

44. Paul M. Schwartz and Karl-Nikolaus Peifer, "Prosser's 'Privacy' and the German Right of Personality: Are Four Privacy Torts Better than One Unitary Concept?," *California Law Review* 98, no. 6 (2010): 1927.

45. Schwartz and Peifer, 1929–1937.

46. Although the Massachusetts Superior Court does not use the term "newsworthy," in summarizing the case, a US District Court in Massachusetts uses the term.

47. Schwartz and Peifer, "Prosser's 'Privacy,'" 1937–1938 (citing Federal Constitutional court, 62 NJW 3089, 3090 (2009)).

48. Sarah E. Igo, "Me and My Data," *Historical Studies in the Natural Sciences* 48, no. 5 (2018): 616–626, https://doi.org/10.1525/hsns.2018.48.5.616.

49. Igo, 618–620.

50. Alan Westin, *Privacy and Freedom* (New York: Atheneum, 1967), 7.

51. Irwin Altman, "Privacy Regulation," 68. Charles Fried also defined privacy in 1968 as "control we have over information about ourselves." Charles Fried, "Privacy," *Yale Law Journal* 77, no. 3 (1968): 475, 477, https://doi.org/10.2307/794941.

52. US Department of Health, Education, and Welfare, *"Records, Computers and the Rights of Citizens," Report of the Secretary's Advisory Committee on Automated Personal Data Systems (HEW Report)*, Washington, DC: July 1973, xx.

53. Colin J. Bennett, *Regulating Privacy: Data Protection and Public Policy in Europe and the United States* (Ithaca, NY: Cornell University Press, 1992), 154–155.

54. Gloria González Fuster, *The Emergence of Personal Data Protection as a Fundamental Right of the EU* (Cham, Switzerland: Springer Science & Business, 2014).

55. González Fuster, 191–193.

56. Orla Lynskey, *The Foundations of EU Data Protection Law* (New York: Oxford University Press, 2015), 195–196.

57. Jef Ausloos, *The Right to Erasure in EU Data Protection Law: From Individual Rights to Effective Protection* (New York: Oxford University Press, 2020), 61.

58. Anthony Giddens, "Living in a Post-Traditional Society," in *Reflexive Modernization: Politics, Tradition and Aesthetics in the Modern Social Order*, ed. Ulrich Beck, Anthony Giddens, and Scott Lash (Redwood City, CA: Stanford University Press, 1994), 74; Nikolas Rose and Peter Miller, *Governing the Present: Administering Economic, Social and Personal Life* (Cambridge, UK: Polity, 2008).

59. Renata Salecl, *The Tyranny of Choice* (London: Profile Books, 2011).

60. Sophia Rosenfeld, "Free to Choose?," *Nation*, June 3, 2014, https://www.thenation.com/article/archive/free-choose/.

61. Radin, *Boilerplate*, 197.

62. *Nudge* references and builds on decades of sociology and behavioral economics including Pierre Bourdieu, *The Social Structures of the Economy* (Cambridge, UK: Polity, 2005), which argued that we are not "individual choosers" but "collective individuals," and Eva Illouz, *Why Love Hurts: A Sociological Explanation* (Cambridge, UK: Polity, 2012), which argued that choice exists in an "ecology of choices." Richard H. Thaler and Cass R. Sunstein, *Nudge: Improving Decisions about Health, Wealth, and Happiness* (New Haven, CT: Yale University Press, 2008).

63. Dan M. Kotliar, "Who Gets to Choose? On the Socio-Algorithmic Construction of Choice," *Science, Technology, & Human Values* 46, no. 2 (2020): 346–375, https://doi.org/10.1177/0162243920925147.

64. Kotliar, 364.

65. Brett Frischmann and Evan Selinger, *Re-Engineering Humanity* (Cambridge, UK: Cambridge University Press, 2018).

66. Meg Leta Jones, "Does Technology Drive Law? The Dilemma of Technological Exceptionalism in Cyberlaw," *Journal of Law, Technology & Policy* (2018): 249–284.

67. Julie E. Cohen, *Between Truth and Power: The Legal Constructions of Informational Capitalism* (New York: Oxford University Press, 2019), 1–2.

68. Cohen, 2; emphasis in original.

69. Sheila Jasanoff, *Designs on Nature: Science and Democracy in Europe and the United States* (Princeton, NJ: Princeton University Press, 2007), 13.

70. Sheila Jasanoff, "Co-Production," SheilaJasanoff.org, accessed August 25, 2023, https://sheilajasanoff.org/research/co-production/.

71. Jasanoff, *Designs on Nature*, 15.

72. James Q. Whitman, "Enforcing Civility and Respect: Three Societies," *Yale Law Journal* 109, no. 6 (2000): 1279, 1282, https://doi.org/10.2307/797466.

73. James Q. Whitman, "Consumerism versus Producerism: A Study in Comparative Law," *Yale Law Journal* 117, no. 3 (2007): 399, https://doi.org/10.2307/20455797.

74. Ann Mettler, "Meet the European Consumer," *Wall Street Journal Europe*, January 29, 2007, https://www.wsj.com/articles/SB117002301171790509; Whitman, "Consumerism versus Producerism," 340.

75. James Q. Whitman, "The Two Western Cultures of Privacy: Dignity versus Liberty," *Yale Law Journal* 113, no. 6 (2004): 1151–1221, https://doi.org/10.2307/4135723.

76. David Vogel, *The Politics of Precaution: Regulating Health, Safety, and Environmental Risks in Europe and the United States* (Princeton, NJ: Princeton University Press, 2012).

77. Vogel, 3

78. Sebastiaan Princen, *EU Regulation and Transatlantic Trade* (Alphen aan den Rijn, Netherlands: Kluwer Law International, 2002); Ernst-Ulrich Petersmann and Mark A. Pollack, eds., *Transatlantic Economic Disputes: The EU, the US, and the WTO* (New York: Oxford University Press, 2003); Vogel, *The Politics of Precaution*, 7.

79. Sheila Jasanoff, "American Exceptionalism and the Political Acknowledgment of Risk," *Daedalus* 119, no. 4 (1990): 63.

80. Jamal Shahin, "A European History of the Internet," *Science and Public Policy* 33, no. 9 (2006): 681–693, https://doi.org/10.3152/147154306781778588; Romain Badouard and Valérie Schafer, "Internet: A Political Issue for Europe (1970s–2010s)," in *Transforming Politics and Policy in the Digital Age*, ed. Jonathan Bishop (Hershey, PA: IGI Global, 2014), 69–84; Andrew L. Russell and Valérie Schafer, "In the Shadow of ARPANET and Internet: Louis Pouzin and the Cyclades Network in the 1970s," *Technology and Culture* 55, no. 4 (2014): 880–907, https://doi.org/10.1353/tech.2014.0096; Julien Mailland and Kevin Driscoll, *Minitel: Welcome to the Internet* (Cambridge, MA: MIT Press, 2017); Joy Lisi Rankin, *A People's History of Computing in the United States* (Cambridge, MA: Harvard University Press, 2018).

81. Mar Hicks, *Programmed Inequality: How Britain Discarded Women Technologists and Lost Its Edge in Computing* (Cambridge, MA: MIT Press, 2017).

82. Martin Campbell-Kelly and Daniel D. Garcia-Swartz, *From Mainframes to Smartphones: A History of the International Computer Industry* (Cambridge, MA: Harvard University Press, 2015), 2–6.

83. Including an interesting back and forth between law professors Paul Schwartz and Anita Allen in 2000, both cutting down the concept from different angles. Paul M. Schwartz, "Internet Privacy and the State," *Connecticut Law Review* 32 (2000): 815–859, https://doi.org/10.2139/ssrn.229011; Anita L. Allen, "Privacy-as-Data Control: Conceptual, Practical, and Moral Limits of the Paradigm," *Connecticut Law Review* 32 (2000): 861–875.

84. See e.g., Joseph Turow, *The Daily You: How the New Advertising Industry Is Defining Your Identity and Your Worth* (New Haven, CT: Yale University Press, 2012); Daniel

J. Solove, *The Digital Person: Technology and Privacy in the Information Age* (New York: NYU Press, 2004); Greg Elmer, *Profiling Machines: Mapping the Personal Information Economy* (Cambridge, MA: MIT Press, 2003).

85. See e.g., John Rothchild, "New European Rules May Give US Internet Users True Privacy Choices for the First Time," *Conversation*, June 14, 2018, https://theconversation.com/neweuropean-rules-may-give-us-internet-users-true-privacy-choices-for-the-first-time-97982 ("Like many privacy rules, the GDPR is based on the principles of notice and choice . . . The concept is part of the Fair Information Practice Principles, a set of privacy guidelines first formulated in a 1973 federal report that now form the basis of many privacy regulations in the U.S. and abroad."); David McCabe, "The Sun May Be Setting on the Old Privacy Rulebook," *Axios*, March 8, 2019, https://www.axios.com/sun-sets-old-privacy-rulebook-notice-consent-5642a827-9a86-454a-ae47-54c74691e8ae.html ("Europe is also heavily invested in the notice and consent approach, which forms the backbone of the General Data Protection Regulation that went into effect last year and has become the de facto global standard."); *Protecting Consumer Privacy in the Era of Big Data: Hearing before the Subcommittee on Consumer Protection and Commerce, Committee on Energy and Commerce*, 116th Congress (February 26, 2019) (Chairman Frank Pallone Jr. stated in his opening remarks, "Yet, for the most part, here in the U.S., no rules apply to how companies collect and use our information. Many companies draft privacy policies that provide few protections and are often unread . . . We can no longer rely on a 'notice and consent' system built on such unrealistic and unfair foundations."), https://energycommerce.house.gov/committee-activity/hearings/hearing-on-protecting-consumer-privacy-in-the-era-of-big-data.

86. See e.g., Kate Fazzini, "Europe's Sweeping Privacy Rule Was Supposed to Change the Internet, But So Far It's Mostly Created Frustration for Users, Companies, and Regulators," CNBC, May 5, 2019, https://www.cnbc.com/2019/05/04/gdpr-has-frustrated-users-and-regulators.html (Quoting Odia Kagan, chair of the GDPR compliance program at law firm Fox Rothschild: "In the end, GDPR is all about consent and it's an approach to privacy that is very European.").

87. Cameron F. Kerry, "Why Protecting Privacy Is a Losing Game Today—and How to Change the Game," *Brookings*, July 12, 2018, https://www.brookings.edu/research/why-protecting-privacy-is-a-losing-game-today-and-how-to-change-the-game/.

88. "Data privacy" is found in US scholarship. See e.g., Neil M. Richards, "Why Data Privacy Law Is (Mostly) Constitutional," *William & Mary Law Review* 56 (2014): 1501–1533; Kelly D. Martin and Patrick E. Murphy, "The Role of Data Privacy in Marketing," *Journal of the Academy of Marketing Science* 45, no. 2 (2017): 135–155, https://doi.org/10.1007/s11747-016-0495-4; Matthew Smith, Christian Szongott, Benjamin Henne, and Gabriele von Voigt, "Big Data Privacy Issues in Public Social Media," *2012 6th IEEE International Conference on Digital Ecosystems and Technologies (DEST)* (2012): 1–6, https://doi.org/10.1109/DEST.2012.6227909.

"Data privacy" is found in policymaking documents. See e.g., DATA Privacy Act, H.R. 8749, 116th Congress (2019–2020); Department of Commerce and National Telecommunications & Information Administration, *Consumer Data Privacy in a Networked World: A Framework for Protecting Privacy and Promoting Innovation in the Global Digital Economy*, White House Privacy Report, February 23, 2012, https://obamawhitehouse.archives.gov/sites/default/files/privacy-final.pdf.

And it's found in reporting. See e.g., Makena Kelly, "Senators Roll Out Bipartisan Data Privacy Bill," *The Verge*, May 20, 2021, https://www.theverge.com/2021/5/20/22444515/amy-klobuchar-data-privacy-protection-facebook-state-laws; Brittany De Lea, "Ohio Introduces Data Privacy Legislation That Allows Residents to Have Their Data Deleted," *Fox Business*, July 13, 2021, https://www.foxbusiness.com/technology/ohio-introduces-data-privacy-legislation-that-allows-residents-to-have-their-data-deleted.

89. For representative "takes" on the relationship between the two (and lack of scholarly treatment on the distinction), see responses to tech policy expert Marcia Hofmann's February 26, 2021, Twitter thread: @marciahofmann (Marcia Hofmann) et al., "My privacy peeps! How do you explain the difference between privacy and data protection?," Twitter, February 26, 2021, https://twitter.com/marciahofmann/status/1365511248431865856. Cynthia Dwork and Deirdre Mulligan published one of the few articles within the US context on the subject, highlighting fairness issues as distinct from privacy (Cynthia Dwork and Deirdre K. Mulligan, "It's Not Privacy, and It's Not Fair," *Stanford Law Review Online* 66 [2013]: 35–40). Since an entire subfield of scholars investigating fairness, bias, and other issues have had to navigate around privacy without the legal foundation of data protection.

90. Barack Obama, "President Barack Obama," interview by Kara Swisher, *Recode Decode* (podcast audio), June 22, 2015, https://podcasts.apple.com/us/podcast/president-barack-obama-the-kara-swisher-interview/id1005732702?i=1000345530560.

91. Ro Khanna, "Rep. Ro Khanna," interview by Kara Swisher, *Recode Decode* (podcast audio), August 15, 2019, https://podcasts.apple.com/us/podcast/decoder-with-nilay-patel/id1011668648?i=1000447016994.

92. Michael Sean Mahoney and Thomas Haigh, eds., *Histories of Computing* (Cambridge, MA: Harvard University Press, 2011).

93. Mahoney and Haigh, 50.

94. "Transcript of Zuckerberg's Appearance before House Committee," *Washington Post*, April 11, 2018, https://www.washingtonpost.com/news/the-switch/wp/2018/04/11/transcript-of-zuckerbergs-appearance-before-house-committee/.

95. Margaret O'Mara, *The Code: Silicon Valley and the Remaking of America* (New York: Penguin Books, 2020).

96. "Transcript of Zuckerberg's Appearance before House Committee."

CHAPTER 2

1. Martin Campbell-Kelly and Daniel D. Garcia-Swartz, *From Mainframes to Smartphones: A History of the International Computer Industry* (Cambridge, MA: Harvard University Press, 2015), 54.

2. Martin Campbell-Kelly, William Aspray, Nathan Ensmenger, and Jeffrey R. Yost, *Computer: A History of the Information Machine*, 3rd ed. (Abingdon, UK: Routledge, 2013), 33–34.

3. Powers avoided infringing on Hollerith's patents by adding an automatic card feeder and sorter, which minimized errors and maximized speed.

4. Heidinger had previously licensed Hollerith Tabulating Machine patents and successfully sold IBM products and services across Europe. Heidinger begrudgingly sold Dehomag to IBM after he missed royalty payments.

5. Lars Heide, "From Invention to Production: The Development of Punched-Card Machines by F. R. Bull and K. A. Knutsen 1918–1930," *Annals of the History of Computing* 13, no. 3 (1991): 261–272, http://doi.org/10.1109/MAHC.1991.10024; Pierre E. Mounier-Kuhn, "Bull: A World-Wide Company Born in Europe," *Annals of the History of Computing* 11, no. 4 (1989): 279–297, http://doi.org/10.1109/MAHC.1989.10045.

6. Heide, "From Invention to Production," 261–268.

7. James W. Cortada, *IBM: The Rise and Fall and Reinvention of a Global Icon* (Cambridge, MA: MIT Press, 2019).

8. Cortada, 126.

9. Cortada, 130.

10. Cortada, 142. Tabulation equipment was not commonly used to locate victims; "Technology even for such limited purposes as census tabulations . . . would not have been found in economically less developed German-occupied eastern Europe where the vast majority of Jews lived." "Locating the Victims," *United States Holocaust Memorial Museum*, accessed September 29, 2023, https://encyclopedia.ushmm.org/content/en/article/locating-the-victims. See also, Michael Allen, "Stranger than Science Fiction: Edwin Black, IBM, and the Holocaust," *Technology and Culture* 43, no. 1 (January 2002): 152–153, https://doi.org/10.1353/tech.2002.0003; David Martin Luebke and Sybil Milton, "Locating the Victim: An Overview of Census-Taking, Tabulation Technology, and Persecution in Nazi Germany," *IEEE Annals of the History of Computing* 16, no. 3 (1994): 34–35, https://doi.org/10.1109/MAHC.1994.298418. However, in France, public employees like René Carmille are sometimes called the first ethical hackers, because they prevented the use of their tabulation data to locate victims. See Lars Heide, "Monitoring People: Dynamics and Hazards of Record Management in France, 1935–1944," *Technology and Culture* 45, no. 1 (2004): 80–101, https://doi.org/10.1353/tech.2004.0020; Amanda Davis, "A History of Hacking," *IEEE*

The Institute, March 6, 2015, https://web.archive.org/web/20150313222911/http://theinstitute.ieee.org/technology-focus/technology-history/a-history-of-hacking. Hollerith punched cards were used to manage the skills of living prisoners in 1942, but were not used for extermination practices until mid-1944. The Nazis found their victims by using existing government and organizational records and their own registration practices—and only rarely did either involve data processing machines. See "Locating the Victims," United States Holocaust Memorial Museum, for the former claim and Cortada, *IBM*, 142, for the latter.

11. Silicon Valley lore has it that a long roll of paper tape containing Altair BASIC software written by Bill Gates and Paul Allen was stolen from their hotel suite by an unknown thief who just plucked it out of a cardboard box. The tape ended up in the hands of Dan Sokol, a semiconductor engineer who was an early member of the Homebrew Computer Club—the legendary nest for the personal computer industry. Sokol made fifty copies to share with the club meeting at the Stanford Linear Accelerator Center. When Gates found out his tape had been freely copied and widely distributed, he sent an angry and now famous letter to PC newsletters and magazines. John Markoff, "A Tale of the Tape from the Days When It Was Still Microsoft," *New York Times*, September 18, 2000, sec. C, 1.

12. David Alan Grier, "From the Editor's Desk," *IEEE Annals of the History of Computing* 26, no. 3 (July–September 2004), 2.

13. The group was referred to as the BUNCH throughout the 1970s.

14. Cortada, *IBM*, 283.

15. Stored-program computing was familiar to large mainframe users like the scientists who operated IBM's massive 7090, but to no one else.

16. Before stored programs, operators took boxes of cards from one operation to another and from one machine to another, all of which had predetermined sequences of tasks that required a physical reconfiguration to change.

17. Dag Spicer, "Back to Life: The Story Behind CHM's IBM 1401 Restoration," *Computer History Museum CORE Magazine* (2009), 12.

18. Cortada, *IBM*, 201.

19. "IBM System 360 Mainframe Computer History Archives 1964 SLT, Course # CH-08," Computer History Archives Project, September 16, 2015, https://www.youtube.com/watch?v=V4kyTg9Cw8g (film and photographs courtesy of IBM Archives Unit with minor restoration edits to improve viewing quality).

20. "While electronic computing held a certain high-tech appeal for many corporate executives in the late 1950s and early 1960s, few had any idea how to integrate effectively this expensive, unfamiliar, and often unreliable technology into their existing operations. It was the computer programmers who developed the applications software that transformed the latent power of a general-purpose computer into a specific tool for solving actual real-world problems." Nathan L. Ensmenger,

"Letting the 'Computer Boys' Take Over: Technology and the Politics of Organizational Transformation," *International Review of Social History* 48, no. S11 (2003): 154, https://doi.org/10.1017/S0020859003001305.

21. "IBM 1401: The Mainframe," IBM 100, accessed September 29, 2023, https://www.ibm.com/ibm/history/ibm100/us/en/icons/mainframe/.

22. Campbell-Kelly and Garcia-Swartz, *From Mainframes to Smartphones*.

23. Laureen Kuo, "Plan Calcul: France's National Information Technology Ambition and Instrument of National Independence," *Business History Review* 96, no. 3 (2022): 590–593, doi:10.1017/S0007680521000441.

24. J. Lyons & Co. bakery built the Lyons Electronic Office I, LEO I, based on Cambridge University's EDSAC.

25. James Sumner, "Defiance to Compliance: Visions of the Computer in Postwar Britain," *History and Technology* 30, no. 4 (2014): 323–325, https://doi.org/10.1080/07341512.2015.1008962.

26. Mar Hicks, *Programmed Inequality: How Britain Discarded Women Technologists and Lost Its Edge in Computing* (Cambridge, MA: MIT Press, 2017).

27. James W. Cortada, "Information Technologies in the German Democratic Republic (GDR), 1949–1989," *IEEE Annals of the History of Computing* 34, no. 2 (2012): 36, https://doi.org/10.1109/MAHC.2012.27.

28. Corinna Schlombs, "The 'IBM Family': American Welfare Capitalism, Labor, and Gender in Postwar Germany," *IEEE Annals of the History of Computing* 39, no. 4 (2017): 13, https://doi.org/10.1353/ahc.2017.0028; for commentary on the labor politics of IBM and the American computer industry generally, see Thomas Haigh, "Computing the American Way: Contextualizing the Early US Computer Industry," *IEEE Annals of the History of Computing* 32, no. 2 (2010): 8–20, https://doi.org/10.1109/MAHC.2010.33.

29. Schlombs, "The 'IBM Family,'" 17.

30. Zuse patented the mechanical memory units, but the Z1 could only read instructions from punch tape reader so the program itself was never stored in memory.

31. Kenneth Flamm, *Creating the Computer* (Washington, DC: Brookings Institution, 1988), 162–164; Campbell-Kelly and Garcia-Swartz, *From Mainframes to Smartphones*, 92.

32. The board was initially supposed to purchase US computer equipment after a report conducted the year prior by the Swedish Royal Academy of Engineering Sciences and a temporary procurement committee had expressed a need for computer machinery in the country, but export controls complicated that strategy. Magnus Johansson, "Early Analog Computers in Sweden—with Examples from Chalmers University of Technology and the Swedish Aerospace Industry," *IEEE Annals of the History of Computing* 18, no. 4 (1996): 27–33, doi: 10.1109/85.539913.

33. "Big Blue" being IBM's nickname; Magnus Johansson, "Big Blue Gets Beaten: The Technological and Political Controversy of the First Large Swedish Computerization Project in a Rhetoric of Technology Perspective," *IEEE Annals of the History of Computing* 21, no. 2 (1999): 14–30, https://doi.org/10.1109/85.761791.

34. Dirk de Wit, "The Construction of the Dutch Computer Industry: The Organisational Shaping of Technology," *Business History* 39, no. 3 (1997): 81–104, https://doi.org/10.1080/00076799700000100.

35. "A Visit to Ottawa's Modern Post Office in 1954," CBC Archives, January 27, 2021, https://www.youtube.com/watch?v=p5ms7Zae9S0.

36. John N. Vardalas, *The Computer Revolution in Canada: Building National Technological Competence* (Cambridge, MA: MIT Press, 2001).

37. Cortada, *IBM*, 89.

38. Montgomery Phister Jr., *Data Processing Technology and Economics* (Bedford, MA: Digital Press, 1979), 287–288. ("All of Western Europe" referred to France, the United Kingdom, West Germany, Belgium, Luxembourg, Denmark, Italy, the Netherlands, Spain, Sweden, Switzerland, Finland, Norway, Austria, Greece, Ireland, and Portugal.)

39. Cortada, *IBM*, 196.

40. Organization for Economic Cooperation and Development, *Gaps in Technology: General Report* (OECD: Paris, 1968), 13.

41. Organization for Economic Cooperation and Development, *Gaps in Technology*, 8.

42. Jean-Jacques Servan-Schreiber, *The American Challenge* (New York: Atheneum, 1968), 13, 101; emphasis in original.

43. Servan-Schreiber, 12.

44. Although Nixdorf was an early and important player in smaller computers, the company didn't move into certain markets like the personal computer and was forced to sell to Siemens in 1990.

45. Eleni Kosta, *Consent in European Data Protection Law* (Leiden, Netherlands: Martinus Nijhoff Publishers, 2013), 43 ("The discussion around privacy in Sweden, as it is also illustrated from the bill on which the Swedish Data Act was based, was influenced by the debate over the right to privacy that had already started in the United States."), n. 132 ("The discussion about privacy protection had started in the United States already in the 1960s and some seminal works were written at that time."). See also, Frits Willem Hondius, *Emerging Data Protection in Europe* (Amsterdam: Elsevier, 1975).

46. The preface also mentions Great Britain's commission but does not, one paragraph later, mention its report by name, instead including only these three.

47. Lars Ilshammar, "When Computers Became Dangerous: The Swedish Computer Discourse of the 1960s," *Human IT: Journal for Information Technology Studies as a Human Science* 9, no. 1 (2007): 6.

48. *Privacy, the Census and Federal Questionnaires: Hearing before the Senate Subcommittee on Constitutional Rights, Committee on the Judiciary*, 91st Congress (April 24, 1969) (testimony of Arthur Miller), 195.

49. Rebecca Lemov calls this archive the "database of dreams" in her book about Bert Kaplan, an anthropologist who attempted to create just that, reaching out to other social scientists to contribute their data about dreams. Rebecca Lemov, *Database of Dreams: The Lost Quest to Catalog Humanity* (New Haven, CT: Yale University Press, 2015); Dan Bouk, "The National Data Center and the Rise of the Data Double," *Historical Studies in the Natural Sciences* 48, no. 5 (2018): 627–636, https://doi.org/10.1525/hsns.2018.48.5.627; Christopher Loughnane and William Aspray, "Rethinking the Call for a US National Data Center in the 1960s: Privacy, Social Science Research, and Data Fragmentation Viewed from the Perspective of Contemporary Archival Theory," *Information & Culture* 53, no. 2 (2018): 203–242, https://doi.org/10.7560/IC53204.

50. Social Science Research Council, *Report of the Committee on the Preservation and Use of Economic Data (Ruggles Report)* (Washington, DC: Social Science Research Council, 1965), 24.

51. Office of Statistical Standards, Bureau of the Budget, *Review of Proposal for a National Data Center* by Edgar S. Dunn (Washington, DC: Office of Statistical Standards, Bureau of the Budget, 1965).

52. The 1966 *Kaysen Report* was republished in the *American Statistician* journal in 1969. Carl Kaysen et al., "Report of the Task Force on the Storage of and Access to Government Statistics," *American Statistician* 23, no. 3 (1969): 11–19, https://doi.org/10.1080/00031305.1969.10481843.

53. *Invasions of Privacy Part 1: Hearings before the Senate Subcommittee on Administrative Procedure, Committee on the Judiciary*, 89th Congress (February 18, 1965).

54. *Invasions of Privacy Part 5: Hearings before the Senate Subcommittee on Administrative Procedure, Committee on the Judiciary*, 89th Congress (October 18, 1965), 2387–2405.

55. *The Computer and Invasion of Privacy: Hearings before the Special Subcommittee on Invasion of Privacy, Committee on Government Operations*, 89th Congress (July 26, 1966) (statement of Cornelius E. Gallagher, Congressman, subcommittee chair), 2.

56. *The Computer and Invasion of Privacy* (July 26, 1966) (statement of Vance Packard, sociologist), 11–12.

57. *The Computer and Invasion of Privacy* (July 26, 1966) (testimony of Vance Packard), 17.

58. *The Computer and Invasion of Privacy* (July 28, 1966) (statement of Paul Baran, computer expert, Rand Corporation), 119–120.

59. *Computer Privacy Part 1: Hearings before the Subcommittee on Administrative Practice and Procedure, Committee on the Judiciary*, 90th Congress (March 14, 1967), 78.

60. Arthur R. Miller, "Computers, Data Banks and Individual Privacy: An Overview," *Columbia Human Rights Law Review* 4, no. 1 (1972): 1–12.

61. HEW was a cabinet-level department from 1953 until 1979, when a separate Department of Education was created and HEW became the Department of Health and Human Services; Carole Parsons, in discussion with the author, September 23, 2020.

62. A great debt is owed to Chris Hoofnagle and Berkeley for digitizing, organizing, and annotating transcripts of the HEW Committee meetings, which had previously not been available. Chris Jay Hoofnagle, "The Origin of Fair Information Practices," *Berkeley Center for Law & Technology*, accessed September 29, 2023, https://www.law.berkeley.edu/research/bclt/research/privacy-at-bclt/archive-of-the-meetings-of-the-secretarys-advisory-committee-on-automated-personal-data-systems-sacapds/.

63. Parsons worked at the National Academy of Sciences national research facility as a typist with the aeronautical sciences group on space projects funded by DARPA in the 1950s.

64. Weizenbaum asked about the term "pre-delinquent," and Lanphere answered, "It's a child that the court calls us and says, 'I'm not going to adjudicate them a delinquent, but I feel like he's on the road, Pat, but if you'll get out there and work with him and get him to quit sniffing glue, or you know, trying to steal a car or something.' It's a child we feel is in danger of becoming a delinquent." Transcript of Proceedings: Meetings of the Secretary's Advisory Committee on Automated Personal Data Systems (April 17, 1972), 225–237.

65. Transcript of Proceedings (April 17, 1972), 230.

66. Transcript of Proceedings (April 17, 1972), 230.

67. Transcript of Proceedings (April 17, 1972), 231.

68. Transcript of Proceedings (April 17, 1972), 233.

69. Stan Aronoff articulated a shared sentiment, "It seemed to me that the people that are best able to articulate the problem and are most afraid are the people who, themselves, work with the computer or are sophisticated in it." Transcript of Proceedings: Meetings of the Secretary's Advisory Committee on Automated Personal Data Systems (May 18, 1972), 96.

70. Transcript of Proceedings: Meetings of the Secretary's Advisory Committee on Automated Personal Data Systems (April 18, 1972); Transcript of Proceedings (May 18, 1972).

71. Transcript of Proceedings: Meetings of the Secretary's Advisory Committee on Automated Personal Data Systems (June 15, 1972), 129.

72. Transcript of Proceedings: Meetings of the Secretary's Advisory Committee on Automated Personal Data Systems (August 17, 1972), 16–17.

73. Transcript of Proceedings (August 17, 1972), 61–63. STS scholars may find Ware's remarks about bridges as hackle-raising, easily reaching for another foundational example of the politics of technology in STS produced by Langdon Winner. In his piece "Do Artifacts Have Politics?," Winner describes how bridges on Long Island were designed to achieve a particular social effect. "Poor people and blacks, who normally used public transit, were kept off the roads because the twelve-foot tall buses could not get through the overpasses. One consequence was to limit access . . . [to a] widely acclaimed public park." Langdon Winner, "Do Artifacts Have Politics?," *Daedalus* 109, no. 1 (Winter 1980), 124.

74. Transcript of Proceedings (June 15, 1972), 146–147.

75. Transcript of Proceedings (April 17, 1972), 210.

76. The Kingston Conference on Computers held in 1970 was cosponsored by federal departments of justice and communications and the Canadian Information Processing Society. Transcript of Proceedings (April 17, 1972), 193, 195.

77. Transcript of Proceedings: Meetings of the Secretary's Advisory Committee on Automated Personal Data Systems (July 25, 1972), 194–196.

78. Memo to Members of the Secretary's Advisory Committee on Automated Personal Data Systems, from David B. H. Martin, Executive Director (June 7, 1972); Willis H. Ware, *RAND and the Information Evolution: A History in Essays and Vignettes* (Santa Monica, CA: RAND, 2008), 157; Robert Gellman, "Fair Information Practices: A Basic History (version 2.20, January 26, 2021)," *BobGellman.com*, updated April 6, 2022, n. 3, https://bobgellman.com/rg-docs/rg-FIPShistory.pdf.; Hoofnagle notes also that Abraham S. Goldstein seems to be a source for elements of the fair information practices: Abraham S. Goldstein, "Legal Control of the Dossier" in *On Record: Files and Dossiers in American Life*, ed. Stanton Wheeler (New York: Russell Sage Foundation, 1969), 415–444.

79. US Department of Health, Education, and Welfare, *"Records, Computers and the Rights of Citizens," Report of the Secretary's Advisory Committee on Automated Personal Data Systems (HEW Report)*, Washington, DC: July 1973, xx–xxi.

80. Memo to Members of the Secretary's Advisory Committee on Automated Personal Data Systems, from David B. H. Martin, Executive Director (June 7, 1972).

81. US Department of Health, Education, and Welfare, *HEW Report*, 48–77.

82. Cortada, *IBM*, 157.

83. US Department of Health, Education, and Welfare, *HEW Report*, 28.

84. *HEW Report*, 22–23.

85. *HEW Report*. "Consumer" is used exclusively throughout with reference to the Fair Credit Reporting Act.

86. Transcript of Proceedings: Meetings of the Secretary's Advisory Committee on Automated Personal Data Systems (September 28, 1972 [PM]), 319–321.

87. *HEW Report*, v–vii..

88. It also states, "The Code should give individuals the right to bring suits for unfair information practices to recover actual, liquidated, and punitive damages, in individual or class actions. It should also provide for recovery of reasonable attorneys' fees and other costs of litigation incurred by individuals who bring successful suits." *HEW Report*, xxiii.

89. Jon Agar, *The Government Machine: A Revolutionary History of the Computer* (Cambridge, MA: MIT Press, 2003).

90. "Robots Will Set Us Free to Serve," *London Times*, October 30, 1965, 9.

91. "Gods from the Machines," *London Times*, July 23, 1947, 5.

92. Walter F. Pratt, *Privacy in Britain* (Lewisburg, PA: Bucknell University Press, 1979), 155–156.

93. Pratt, 19–37.

94. Lord Windlesham introduced a motion to call attention to computer stored personal information in 1969, clarifying, "To me, the guiding principle should surely be that the individual has a right to know of and to limit the circulation of information about himself . . . It is in this sense an ingredient of the individual's right to privacy and one that should not be infringed without the showing of an overriding social need and that protective safeguards have been established in the first instance and satisfied thereafter." 306 Hansard, 5th series, H.L. col. 103 (December 3, 1969).

95. Kenneth Baker's Data Surveillance Bill 1969 provided for a registration and printout for everyone listed in the computer database and Leslie Huckfield's Control of Personal Information Bill 1971 provided for a licensing tribunal. Secretary of State for the Home Department, *Report of the Committee on Privacy (Younger Report)*, July 1972, 186–189, par. 604–612.

96. *Younger Report*, 1.

97. *Younger Report*, 10 par. 38; see also chapter 4.

98. *Younger Report*, 191.

99. Ware, *RAND and the Information Evolution*, 157.

100. Seasoned privacy professional Bob Gellman has for years been editing and updating a history of the FIPs posted online at his personal website. Robert Gellman, "Fair Information Practices: A Basic History (version 2.22, April 6, 2022)," *BobGellman.com*, updated April 6, 2022, https://bobgellman.com/rg-docs/rg-FIPShistory.pdf.

101. British Computer Society memo Reference PRI/70 4/2/19, HO 261—Data Protection Committee, Box 264-121, File 3. British National Archives Kew.

102. BCS oral testimony, July 15, 1971, HO 261—Data Protection Committee, Box 264-121, File 20.

103. *Younger Report*, appendix N, 332.

104. D'Agapeyeff was most famous for his 1939 cryptography book *Codes and Ciphers*, which included "challenge cipher" at the end that was never solved and d'Agapeyeff forgot how he encrypted it. Alexander d'Agapeyeff, *Codes and Ciphers: A History of Cryptography* (New York: Read Books, 2016); CAP built compilers and system software before expanding and merging with the French company Sema-Metra in 1988.

105. BCS evidence, HO 261—Data Protection Committee, Box 264-121, File 53.

106. BCS oral testimony, July 15, 1971, HO 261—Data Protection Committee, Box 264-121, File 22.

107. BCS evidence, prepared for internal subcommittee May 6, 1971, HO 261—Data Protection Committee, Box 264-121, File 73.

108. BCS oral testimony, July 15, 1971, HO 261—Data Protection Committee, Box 264-121, File 224.

109. BCS evidence, HO 261—Data Protection Committee, Box 264-121, File 58.

110. BCS Code of Conduct Appendix A, Notes and Guidance 2.12, HO 261—Data Protection Committee, Box 264-121, Files 48-50.

111. The DPMA established a code of ethics for the profession at its founding in 1951, making enhancements over the years. "Brief History of AFIPS and Its Constituent Societies," in *Annals of the History of Computing* 8, no. 3 (1986): 219-224.

112. Donn B. Parker, "Rules of Ethics in Information Processing," *Communications of the ACM* 11, no. 3 (1968): 198-201, https://doi.org/10.1145/362929.362987; ACM Ethics, "Historical Archive of the ACM Code of Ethics," accessed December 18, 2023, https://ethics.acm.org/code-of-ethics/previous-versions/.

113. The ten principles were:

(i) The purpose of holding data should be specified.

(ii) There should only be authorized access to data.

(iii) There should be minimum holdings of data for specified purposes.

(iv) Persons in statistical surveys should not be identified.

(v) Subject access to data should be given.

(vi) There should be security precautions for data.

(vii) There should be security procedures for personal data.

(viii) Data should only be held for limited relevant periods.

(ix) Data should be accurate and up to date.

(x) Any value judgements should be coded.

114. *Younger Report*, 189.

115. *Younger Report*, 178.

116. *Younger Report*, 191 ("We do not believe that the time is ripe for the sort of detailed controls advocated in the Bills proposed by Mr Baker and Mr Huckfield, though some scheme of registration, licensing and inspection on these lines may be appropriate at a future date.").

117. Transcript of Proceedings (August 17, 1972), 56–67; *Younger Report*, 191.

118. Computers and Privacy (Cmnd 6353), HMSO, London (1975), 8.

119. Adam Warren and James Dearnley, "Data Protection Legislation in the United Kingdom: From Development to Statute 1969–84," *Information, Communication & Society* 8, no. 2 (2005): 238–263, https://doi.org/10.1080/13691180500146383.

120. Warren and Dearnley.

121. Ilshammar, "When Computers Became Dangerous," 11.

122. Ilshammar, 9.

123. In 1969 the OSK was tasked with the public sector while the Credit Information Commission assessed the need for the privacy sector, but a new mandate in May 1971 directed the OSK to look at a comprehensive data law.

124. Ilshammar, "When Computers Became Dangerous," 24.

125. David H. Flaherty, *Protecting Privacy in Surveillance Societies: The Federal Republic of Germany, Sweden, France, Canada, and the United States* (1992; ebook, Chapel Hill: University of North Carolina Press, 2014), 94.

126. Chantal Lebrument and Fabien Soyez, *The Inventions of Louis Pouzin: One of the Fathers of the Internet* (Cham, Switzerland: Springer International, 2020).

127. Not coincidentally, according to Georges, the name was chosen by the engineer working on the project housed within the Ministry of the Economy and Finance, at that time led by Valéry Giscard d'Estaing, who was heading off to Africa for a safari. Marie Georges, in discussion with the author, August 19, 2019.

128. Philippe Boucher, "'Safari' Ou la Chasse aux Français," *Le Monde*, March 21, 1974, 9; André Vitalis, *Informatique, Pouvoir et Libertés (2e éd.)* (Economica, 1988), 77–90; "De Safari à Edvige: 35 années d'une Histoire oubliée malgré la création de la CNIL," Mag-Securs 21 (2008), reprinted online http://www.mag-securs.com/news/articletype/articleview/articleid/23700/de-safari-a-edvige--35-annees-d8217une-histoire-oubliee-malgre-la-creation-de-la-cnil.aspx.

129. *Rapport de la Commission Informatique Libertés (Tricot Report)*, Paris: 1975; *Report of the Committee Informatics and Liberties* (Council of Europe translation), Strasbourg: June 10, 1976.

130. *Tricot Report*, 7.

131. Françoise Gallouédec-Genuys and Herbert Maisl, *Le Secret des Fichiers* (Paris: Cujas, 1976). For coverage of the book see, "'Hunt the Frenchman' Bill Deferred as Privacy Becomes Issue in Paris," *New Scientist* 71, no. 1020 (September 30, 1976): 698; Flaherty, *Protecting Privacy in Surveillance Societies*.

132. *Tricot Report*, 7.

133. Flaherty, *Protecting Privacy in Surveillance Societies*, 167.

134. Flaherty, 179–181.

135. Simon Nora and Alain Minc, *Computerization of Society: A Report to the President of France* (Cambridge, MA: MIT Press, 1980).

136. Confédération française démocratique du travail, *Les degats du progres: Les travailleurs face au changement technique* (Paris: Éditions du Seuil, 1977) (The book was written in collaboration with Jean Louis Missika, Dominique Wolton, and the Confédération française démocratique du travail [French Democratic Confederation of Labour]).

137. Richardson came to be called the Watergate Martyr.

138. US Privacy Protection Study Commission, *Personal Privacy in an Information Society (PPSC Report)*, 1977, preface.

139. US Privacy Protection Study Commission, *PPSC Report*, epilogue.

140. Including Tennessen, who met with Minnesotan companies but had never at that point seen a computer. Interview with Robert Tennessen.

141. US Privacy Protection Study Commission, *PPSC Report*, introduction.

142. US Privacy Protection Study Commission, *PPSC Report*, epilogue.

143. US Privacy Protection Study Commission, *PPSC Report*, introduction.

144. Colin J. Bennett, *Regulating Privacy: Data Protection and Public Policy in Europe and the United States* (Ithaca, NY: Cornell University Press, 1992); Colin J. Bennett and Charles D. Raab, *The Governance of Privacy: Policy Instruments in Global Perspective* (Cambridge, MA: MIT Press, 2006); Colin J. Bennett and Charles D. Raab, "Revisiting the Governance of Privacy: Contemporary Policy Instruments in Global Perspective," *Regulation & Governance* 14, no. 3 (2018): 447–464, https://doi.org/10.1111/rego.12222.

145. *Hew Report*, 169.

146. Simitis Spiros, "From the Market to the Polis: The EU Directive on the Protection of Personal Data," *Iowa Law Review* 80, no. 3 (1995): 447.

147. Gesetz zur Fortentwicklung der Datenverarbeitung und des Datenschutzes (Bundesdatenschutzgesetz) vom 20 Dezember 1990 (December 20, 1990), reprinted in *Data Protection in the European Community: The Statutory Provisions*, ed. Spiros Simitis et al. (Baden-Baden, Germany: Nomos, 1994).

148. Paul M. Schwartz, "European Data Protection Law and Restrictions on International Data Flows," *Iowa Law Review* 80, no. 3 (1995): 475.

149. Gesetz zum Schutz vor Mißbrauch personenbezogener Daten bei der Datenverarbeitung (Bundesdatenschutzgesetz or BDSG) (Law to Protect Against Misuse of Personal Data in Data Processing [Federal Data Protection Law]), Jan. 27, 1977 § 2(1); J. Lee Riccardi, "The German Federal Data Protection Act of 1977: Protecting the Right to Privacy?," *Boston College International & Comparative Law Review* 6, no. 1 (1983): 249.

150. For a comparison between US and German constitutional responses to data processing over the same period see Paul M. Schwartz, "The Computer in German and American Constitutional Law: Towards an American Right of Informational Self-Determination," *American Journal of Comparative Law* 37, no. 4 (1989): 675–701, https://doi.org/10.2307/840221.

151. Schwartz, 688 (citing "Volkszihlung: Lasst 1000 Fragebogen gliuhen," *Spiegel* Nr. 13, 28 (1983)); Larry Frohman, *The Politics of Personal Information: Surveillance, Privacy, and Power in West Germany* (New York: Berghahn Books, 2020), 137.

152. Schwartz, "The Computer in German and American Constitutional Law," 691.

153. Schwartz, 690.

154. Frohman, *The Politics of Personal Information*.

155. Flaherty, *Protecting Privacy in Surveillance Societies*, 96.

156. Flaherty, 104.

157. Charles K. Wilk, ed., Selected Foreign National Data Protection Laws & Bills, Department of Commerce March 1978, 70.

158. Flaherty, *Protecting Privacy in Surveillance Societies*, 139 (citing Svenska Dagbladet, January 29, 1983).

159. Flaherty, 95.

160. Hans Corell, "Technological Development and Its Consequence for Data Protection," in *Beyond 1984: The Law and Information Technology in Tomorrow's Society* (Strasbourg, France: Council of Europe, 1985), 59.

161. Privacy Act of 1974 and Amendments Sec. 552a(2).

162. S. Rep. No. 1183, 93d Cong., 2d Sess. 79 (1974).

163. Michael Schudson, *The Rise of the Right to Know: Politics and the Culture of Transparency, 1945–1975* (Cambridge, MA: Harvard University Press, 2015), 63.

164. Privacy Act of 1974, 5 U.S.C. § 552a(b)(3); emphasis added.

165. Privacy Act, 5 U.S.C. § 552a(j)(1) and (j)(2).

166. US Privacy Protection Study Commission, *PPSC Report*, "Appendix 4: The Privacy Act of 1974: An Assessment."

167. David F. Linowes, "The US Privacy Protection Commission: A Retrospective View from the Chair," *American Behavioral Scientist* 26, no. 5 (1983): 577–590, https://doi.org/10.1177/000276483026005005.

168. US Privacy Protection Study Commission, *PPSC Report*, "Chapter 3: The Depository Relationship."

169. *Tricot Report*, 17 ("It is in fact central government, big cities, big business, etc., whose power increases, while every individual feels himself more and more exposed to society, controlled by it, and unable to influence the choices he is invited to participate in.").

170. Flaherty, *Protecting Privacy in Surveillance Societies*, 171 (citing 01-Informatique Hebdo, May 18, 1981); see also Rex Malik, "France's Social Agenda for *Le Computer*," *Computer World* 17, no. 19 (May 9, 1983): 1–22.

171. Philip Faflick and Pam Schirmeister, "A Terminal in Every Home," *Time* 121, no. 11 (September 13, 1982): 65.

172. Julien Mailland and Kevin Driscoll, "Minitel: The Online World France Built Before the Web," *IEEE Spectrum*, June 20, 2017, https://spectrum.ieee.org/minitel-the-online-world-france-built-before-the-web; Julien Mailland and Kevin Driscoll, *Minitel: Welcome to the Internet* (Cambridge, MA: MIT Press, 2017).

173. For a selection of terminal makes and models, see Julien Mailland and Kevin Driscoll, "Terminals," Minitel Research Lab, USA, accessed August 24, 2023, https://www.minitel.us/terminals.

174. Flaherty, *Protecting Privacy in Surveillance Societies*.

175. Privacy Act, R.S.C., 1985, c. P-21.

176. Data Protection Act 1984, Chapter 35, Section 1(4); Data Protection Act 1984, Chapter 35, Section 1(5).

177. Data Protection Act 1984, Chapter 35, Section 32(3)(d).

178. Warren and Dearnley, "Data Protection Legislation in the United Kingdom," 255–256.

179. Schwartz, "European Data Protection Law and Restrictions on International Data Flows," 472.

180. Eric J. Novotny, "Transborder Data Flows and International Law: A Framework for Policy-Oriented Inquiry," *Stanford Journal of International Law* 16 (1980): 143–144.

181. Michael D. Kirby, "Transborder Data Flows and the 'Basic Rules' of Data Privacy," *Stanford Journal of International Law* 16 (1980): 28.

182. Rapporteur Karl Czernetz, *Human Rights and Modern Scientific and Technological Developments*, Council of Europe—Parliamentary Assembly, Doc 2326, 22.01.1968 (1968).

183. Council of Europe, *Resolution (73) 22 on the Protection of the Privacy of Individuals vis-à-vis Electronic Data Banks in the Private Sector*, 1973; Council of Europe, *Resolution 74 (29) on the Protection of the Privacy of Individuals vis-à-vis Electronic Data Banks in the Public Sector*, 1974.

184. Council of Europe, Convention for the Protection of Individuals with Regard to Automatic Processing of Personal Data (ETS. 108) (28.01.1981) [Convention 108].

185. Consent is mentioned in article 15(3), which reads, "In no case may a designated authority be allowed to make under Article 14, paragraph 2, a request for assistance on behalf of a data subject resident abroad, of its own accord and without the express consent of the person concerned."

186. Convention 108, Chapter I, Art 2(a).

187. Frits Hondius, "Data Law in Europe," *Stanford Journal of International Law* 16 (1980): 88–89.

188. Hondius, 110.

189. Organization for Economic Cooperation and Development, *Guidelines Governing the Protection of Privacy and Transborder Flow of Personal Data* (Paris: OECD Publishing, 1980), 3.

190. Organization for Economic Cooperation and Development, "The Evolving Privacy Landscape: 30 Years After the OECD Privacy Guidelines," *OECD Digital Economy Papers* 176 (Paris: OECD Publishing, 2011): 8, https://doi.org/10.1787/5kgf09z90c31-en.

191. See e.g., Herbert Maisl, "Legal Aspects of Data Flows between Public Agencies in France," *Computer Networks (1976)* 3, no. 3 (1979): 199–204, https://doi.org/10.1016/0376-5075(79)90041-2.

192. Priscilla M. Regan, "Personal Information Policies in the United States and Britain: And the Dilemma of Implementation Considerations," *Journal of Public Policy* 4, no. 1 (1984): 19–38; Priscilla M. Regan, *Legislating Privacy: Technology, Social Values, and Public Policy* (Chapel Hill: University of North Carolina Press, 1995).

193. James B. Rule, *Private Lives and Public Surveillance: Social Control in the Computer Age* (New York: Schocken Books, 1974), 13–14.

194. Joel Reidenberg, in discussion with the author, March 22, 2019.

195. Paul Schwartz and Joel Reidenberg, "On-line Services Data Protection Law and Privacy: Regulatory Responses," Official Publication of the European Union, 1998 (study carried out for the Commission of the European Communities (DGXV) regarding online privacy in Belgium, France, Germany, and the United Kingdom); Paul Schwartz and Joel Reidenberg, *Data Privacy Law* (Charlottesville, VA: Michie Publishing/Lexis Law Publishing, 1996).

NOTES

196. Spiros, "From the Market to the Polis."

197. "Communication on the protection of individuals in relation to the processing of personal data in the Community and information security," Commission of the European Communities, COM (90) 314 final—SYN 287—SYN 288 (September 13, 1990).

198. Spiros, "From the Market to the Polis," 447.

199. Proposal for a Council Directive Concerning the Protection of Individuals in Relation to the Processing of Personal Data, 1990 oJ. (C 277) 3, Explanatory Memorandum II; Amended Proposal for a Council Directive on the Protection of Individuals with Regard to the Processing of Personal Data and on the Free Movement of Such Data, 1992 O.J. (C 310) 38, paras. 1–2, 10.

200. Amended Proposal for a Council Directive on the Protection of Individuals with Regard to the Processing of Personal Data and on the Free Movement of Such Data, 1992 O.J. (C 310) 38, recitals para. 9.

201. Spiros, "From the Market to the Polis," 449.

202. Artikel 3(1) ("'personenbezogene Daten' alle Informationen, die sich auf eine identifizierte oder identifizierbare natürliche Person (im Folgenden 'betroffene Person')") https://eur-lex.europa.eu/legal-content/DE/TXT/PDF/?uri=CELEX:32016L0680&from=EN; Article 3(1) ("'données à caractère personnel,' toute information se rapportant à une personne physique identifiée ou identifiable [ci-après dénommée 'personne concernée']"), https://eur-lex.europa.eu/legal-content/FR/TXT/PDF/?uri=CELEX:32016L0680&from=EN; Article 3(1) ("'personal data' means any information relating to an identified or identifiable natural person ['data subject']"), https://eur-lex.europa.eu/legal-content/EN/TXT/PDF/?uri=CELEX:32016L0680&from=EN.

203. Marie Georges, in discussion and emails with the author, August 19, 2019.

204. Georges, August 19, 2019.

205. Alex Fitzpatrick, Lisa Eadicicco, and Matt Peckham, "The 15 Most Influential Websites of All Time," *Time*, October 20, 2017, http://time.com/4960202/most-influential-websites/; W. Joseph Campbell, *1995: The Year the Future Began* (Berkeley: University of California Press, 2015).

206. Jeffrey A. Hart, Robert R. Reed, and François Bar, "The Building of the Internet: Implications for the Future of Broadband Networks," *Telecommunications Policy* 16, no. 8 (1992): 666–689, https://doi.org/10.1016/0308-5961(92)90061-S.

207. Richard Brandt and Amy Cortese, "Bill Gates's Vision: He's Pushing Microsoft Past the PC and onto the Info Highway," *Business Week*, June 27, 1994, 56, 60 ("These international transmissions of data depend on a variety of digital world networks, both wired and wireless. According to a recent report, Bill Gates, the chief executive officer of Microsoft Corp., believes that the future of the computer industry rests with these networks.").

208. David E. Sanger, "Bailing Out of the Mainframe Industry," *New York Times*, February 5, 1984, sec. 3, 1.

209. Campbell-Kelly and Garcia-Swartz, *From Mainframes to Smartphones*.

210. Ruth Walker, "In Europe, Microcomputer Industry Is Making Up for Lost Time," *Christian Science Monitor*, October 4, 1985, available at https://www.csmonitor.com/1985/1004/fmicro.html.

211. Walker.

212. Walker.

213. Hondius, "Data Law in Europe," 89 ("Data protection represents an example of the third type of legal response to computer problems—creation of an entirely new body of law.").

CHAPTER 3

1. John S. Quarterman and Josiah C. Hoskins, "Notable Computer Networks," *Communications of the ACM* 29, no. 10 (1986): 932–971, https://doi.org/10.1145/6617.6618; Kevin Driscoll and Camille Paloque-Berges, "Searching for Missing 'Net Histories,'" *Internet Histories* 1, nos. 1–2 (2017): 47–59, https://doi.org/10.1080/24701475.2017.1307541.

2. The two classic texts on these notably different threads in the history of computing are Paul N. Edwards, *The Closed World: Computers and the Politics of Discourse in Cold War America* (Cambridge, MA: MIT Press, 1996); and Fred Turner, *From Counterculture to Cyberculture: Stewart Brand, the Whole Earth Network, and the Rise of Digital Utopianism* (Chicago: University of Chicago Press, 2006).

3. Joanne McNeil, *Lurking: How a Person Became a User* (New York: MCD, 2020), 6–8. Computer historians Laine Nooney, Kevin Driscoll, and Kera Allen find empirical evidence of this departure in the early pages of *Softalk* magazine, a publication dedicated to the Apple II: "In the 1980s, the user emerged as a distinct class of personal computer owner motivated by instrumental goals rather than the exploratory pleasures of hackers and hobbyists" (Laine Nooney, Kevin Driscoll, and Kera Allen, "From Programming to Products: *Softalk* Magazine and the Rise of the Personal Computer User," *Information & Culture* 55, no. 2 (2020): 105–129.).

4. Gabriella Coleman, *Hacker, Hoaxer, Whistleblower, Spy: The Many Faces of Anonymous* (New York: Verso Books, 2014); Gabriella Coleman, *Coding Freedom: The Ethics and Aesthetics of Hacking* (Princeton, NJ: Princeton University Press, 2013).

5. "User," The Jargon File (version 4.4.7), accessed September 29, 2023, http://www.catb.org/~esr/jargon/html/U/user.html.

6. "Luser," The Jargon File, accessed September 29, 2023, http://www.catb.org/~esr/jargon/html/L/luser.html.

7. Philip L. Frana, "Telematics and the Early History of International Digital Information Flows," *IEEE Annals of the History of Computing* 40, no. 2 (2018): 32–47, 34, https://doi.org/10.1109/MAHC.2018.022921442 ("Telematics liberated data for new export markets. It permitted the automation of work, especially remote work done at a distance. It encouraged the international division of labor. It inspired specialization and centralization of functions, including aggregation and storage, and localization of data consumption.").

8. Klaus Larres, *Uncertain Allies: Nixon, Kissinger, and the Threat of a United Europe* (New Haven, CT: Yale University Press, 2021).

9. Patrice Flichy, *Dynamics of Modern Communication: The Shaping and Impact of New Communication Technologies* (Thousand Oaks, CA: SAGE Publications, 1995); Eli Noam, *Telecommunications in Europe* (New York: Oxford University Press, 1992).

10. Challenging the accepted pre-allocation techniques in telecommunications at the time was driven by J. C. R. Licklider's direction of the Advanced Research Projects Agency (ARPA) over 1962–1964, when he sponsored and supported the development of time-sharing computer systems and rethinking communication between computers. Lawrence G. Roberts, "The Evolution of Packet Switching," *Proceedings of the IEEE* 66, no. 11 (1978): 1307–1313, http://www.ece.ucf.edu/~yuksem/teaching/nae/reading/1978-roberts.pdf.

11. Janet Abbate, *Inventing the Internet* (Cambridge, MA: MIT Press, 2000).

12. The effort was supported by the British Post Office, the British Library, the British Ministry of Defense, the Department of Labor, and the Science Research Council, while other efforts were pursued as well. Peter T. Kirstein, "Early Experiences with the ARPANET and Internet in the United Kingdom," *IEEE Annals of the History of Computing* 21, no. 1 (1999): 38–44, https://doi.org/10.1109/85.759368.

13. Sandra Braman, "Privacy by Design: Networked Computing, 1969–1979," *New Media & Society* 14, no. 5 (2012): 798–814, https://doi.org/10.1177/1461444811426741.

14. Rita Zajácz, "WikiLeaks and the Problem of Anonymity: A Network Control Perspective," *Media, Culture & Society* 35, no. 4 (2013): 489–505, https://doi.org/10.1177/0163443713483793.

15. Jamal Shahin, "A European History of the Internet," *Science and Public Policy* 33, no. 9 (2006): 681, https://doi.org/10.3152/147154306781778588.

16. Noam, *Telecommunications in Europe*.

17. For instance, in France the PTT was not helpful to the Cyclades team developing the datagrams because it had already started using its own.

18. Andrew L. Russell, "The Internet That Wasn't," *IEEE Spectrum* 50, no. 8 (2013): 40, https://doi.org/10.1109/MSPEC.2013.6565559.

19. Roberts, "The Evolution of Packet Switching."

20. The RCP network could, explained its designer Rémi Després to Andrew Russell and Valérie Schafer in a 2012 interview, "give the guarantee of never losing anything except when there is hardware failure or link failure." Andrew L. Russell and Valérie Schafer, "In the Shadow of ARPANET and Internet: Louis Pouzin and the Cyclades Network in the 1970s," *Technology and Culture* 55, no. 4 (2014): 880–907, 897, https://doi.org/10.1353/tech.2014.0096.

21. Chantal Lebrument and Fabien Soyez, *The Inventions of Louis Pouzin: One of the Fathers of the Internet* (Cham, Switzerland: Springer International, 2020).

22. Russell, "The Internet That Wasn't," 40.

23. Russell and Schafer, "In the Shadow of ARPANET and Internet," 899–900.

24. Russell and Schafer, "In the Shadow of ARPANET and Internet," 887.

25. Dorian James Rutter, "From Diversity to Convergence: British Computer Networks and the Internet, 1970–1995" (PhD diss., University of Warwick, 2005); Shirley F. Redpath, *With All Due Respect: A History of the Real Time Club* (2013), available at http://realtimeclub.co.uk/wp-content/uploads/With-All-Due-Respect.pdf.

26. J. Howlett, *Report of the National Committee on Computer Networks* (London: Department of Industry, 1978).

27. Kirstein, "Early Experiences with the ARPANET and Internet in the United Kingdom."

28. Walter F. Pratt, *Privacy in Britain* (Lewisburg, PA: Bucknell University Press, 1979); David G. Barnum, "Judicial Oversight of Interception of Communications in the United Kingdom: An Historical and Comparative Analysis," *Georgia Journal of International and Comparative Law* 44, no. 2 (2016): 237–304.

29. Pratt, *Privacy in Britain*, 121 (citing *News Chronicle*, June 7, 1957, 4.)

30. Kenneth Ellis, *The Post Office in the Eighteenth Century: A Study in Administrative History* (New York: Oxford University Press, 1958).

31. Edward Raymond Turner, "The Secrecy of the Post," *English Historical Review* (1918): 320–327.

32. Committee of Privy Councillors, *Inquiry into the Interception of Communications, Birkett Report*, 1957, Cmnd. 283, available at http://www.fipr.org/rip/Birkett.htm.

33. *Birkett Report*; Julie M. Flavell, "Government Interception of Letters from America and the Quest for Colonial Opinion in 1775," *William and Mary Quarterly* 58, no. 2 (2001): 403–430, https://doi.org/10.2307/2674191.

34. *Birkett Report*.

35. Prior to 1998, violations could be alleged to the European Commission, which then decided whether or not to refer cases to the European Court of Human Rights.

36. Interception of Communications Act 1985, c. 56, § 1(1).

37. Interception of Communications Act 1985, c. 56, § 1(2).

38. Computer Misuse Act 1990, c. 18.

39. The two met on bulletin board systems, where they also met teenager Paul Bedworth. Bedworth was acquitted despite admitting to the conduct, but insisted it was an obsession. Bedworth was only fourteen at the time, while Strickland and Woods were in their early twenties. Charles Arthur, "Hacker's Acquittal Casts Doubt on Law," *New Scientist*, March 27, 1993, https://www.newscientist.com/article/mg13718660-300-hackers-acquittal-casts-doubt-on-law/.

40. Duncan Campbell, "From the Archive, 22 May 1993: British Computer Hackers Behind Bars," *The Guardian*, May 22, 2013, https://www.theguardian.com/theguardian/2013/may/22/hackers-hacking-computer-jailed.

41. When Prestel failed in the UK, France Telecom actually sold Minitel services back to Britain beginning in 1993. Frank Barrett, "Minitel Gets on the Road: France's Successful Telephone Information and Sales Service Based on Prestel Is Being Made Available in Britain," *Independent*, December 17, 1993, https://www.independent.co.uk/extras/indybest/gadgets-tech/computers-minitel-gets-on-the-road-france-s-successful-telephone-information-and-sales-service-based-on-prestel-is-being-made-available-in-britain-frank-barrett-reports-1468011.html.

42. Canada developed a more sophisticated but similarly arranged videotex (and teletext) system called Telidon through its Communications Research Center of the Canadian Department of Communication, which was a partnership with Bell Canada and TV Ontario. The investments over the 1980s, including those by media firms in the US, supported systems that were defunct by the late 1990s.

43. Julien Mailland and Kevin Driscoll, *Minitel: Welcome to the Internet* (Cambridge, MA: MIT Press, 2017).

44. Julien Mailland and Kevin Driscoll, "The French Connection Machine," *IEEE Spectrum* 54, no. 7 (2017): 32–37.

45. "Users pay only for the time they spend on le Kiosque and are charged on their monthly telephone bills, at a rate of 77 centimes (about 10 cents) for every 45 seconds. Le Kiosque's service providers keep about 60 percent of the fee, with the remainder going to the P.T.T. More than half of the country's Minitel traffic is on le Kiosque." Nadine Epstein, "Et voila! Le Minitel," *New York Times Magazine*, March 9, 1986, 46.

46. Epstein, 46.

47. Mailland and Driscoll, "The French Connection Machine," 37.

48. "Comparative Study on Wiretapping and Electronic Surveillance Laws in Major Foreign Countries," Library of Congress, Law Library (1975).

49. Although most European constitutional codification of the 1800s and early 1900s included explicit communication privacy rights in the form of confidential

correspondence, France has never explicitly codified a right to privacy in its constitutions, which were considered only inspirational until the mid-twentieth century. The drafters of 1789's *La Déclaration des Droits de l'Homme et du Citoyen* (The Declaration of the Rights of Man and of the Citizen) understood communication privacy rights tied so tightly to the freedom of expression that they determined including an additional right to confidential correspondence was redundant in the final drafts. Blanca Ruiz, *Privacy in Telecommunications: A European and an American Approach* (Cham, Switzerland: Springer, 1997). For comparison of constitutional moments in domestic privacy law, see David Erdos, "Comparing Constitutional Privacy and Data Protection Rights within the EU" (University of Cambridge Faculty of Law Research Paper No. 21/2021), https://doi.org/10.2139/ssrn.3843653. Before 1958, the law passed by Parliament could not be overturned by judges, as they lacked any authority to invalidate legislation. In 1958, the Fifth Republic's constitution created a new body called the Constitutional Council, which did have the authority to determine whether legislation conformed to the constitution. But the Council only has a short window to determine unconstitutionality—after it's passed and before it's promulgated. When, in 1974, France ratified the European Convention on Human Rights from 1953, Parliament incorporated article 8 (right to "respect for one's private and family life, his home and his correspondence") into French law. Still, the constitutionality of French law was not challenged by those it has been applied to until a significant 2008 constitutional revision that now allows parties to present *Question Prioritaire de Constitutionnalité* (QPC) to the supreme jurisdiction of the particular area of law, who can then send it to the Constitutional Court for review. This makes for a set of legal privacy events quite distinct from other countries.

50. Judgment of June 12, 1952, Case. Crim., 1952 J.C.P. Jur. No. 7241 note J. Brouchot (the *Imbert* decision).

51. Loi n* 70-643 of July 17, 1970.

52. Edward A. Tomlinson, "The Saga of Wiretapping in France: What It Tells Us about the French Criminal Justice System," *Louisiana Law Review* 53, no. 4 (1992): 1091–1151.

53. Gerard Alberts and Ruth Oldenziel, eds., *Hacking Europe: From Computer Cultures to Demoscenes* (Cham, Switzerland: Springer, 2014).

54. Other translations call it the Committee for the Liquidation and Diversion of Computers. CLODO is consistently the chosen acronym; William Dowell, "French Terrorists Attract Outside Aid," *Christian Science Monitor*, July 17, 1980, https://www.csmonitor.com/1980/0717/071745.html; "International Report," *Computerworld*, November 14, 1983.

55. "Le Clodo revendique l'attentat de Toulouse," *Libération*, April 9, 1980. A number of translations of the letter can be found in various sources including: John Lamb and James Etheridge, "DP: The Terror Target," *Datamation* 32, no. 3 (February 1, 1986): 44–46; August Bequai, *Techno-Crimes: The Computerization of Crime*

and Terror (Washington, DC: Lexington Books, 1987), 129; Edward Moxon-Browne, "Terrorism in France," in *Contemporary Terrorism*, ed. William Gutteridge (New York: Facts on File, 1986), 111–134; "CLODO Communique Following Attack on Philips Data Systems (1980)," *The Anarchist Library*, April 9, 1980, http://theanarchistlibrary.org/library/unknownrevolutions-clodo-communique-following-attack-on-philips-data-systems-1980.

56. Pierre-Alain Weill, "État de la législation et tendances de la jurisprudence relatives à la protection des données personnelles en droit pénal français," *Revue internationale de droit comparé* 39, no. 3 (1987): 655–675.

57. Act no. 88–19 of 5 January 1988 on computer fraud [Godfrain Law or Loi Godfrain], articles 323–1 to 323–7.

58. Ward Christensen and Randy Suess, "Hobbyist Computerized Bulletin Board," *Byte Magazine* 3, no. 11 (1978): 150–158.

59. Michael A. Banks, *On the Way to the Web: The Secret History of the Internet and Its Founders* (Berkeley, CA: Apress, 2008).

60. Judith Berck, "All About Electronic Bulletin Boards: It's No Longer Just Techno-Hobbyists Who Meet by Modem," *New York Times*, July 19, 1992, sec. 3, 12.

61. Brit Hume, "A Primer on Using an Electronic Bulletin Board System," *Washington Post*, February 6, 1989, 31.

62. Kevin Ackermann, "The Old Puppet Masters: Content Moderation on Computer Bulletin Board Systems" (M.A. thesis, Georgetown University, 2020).

63. Martin Lasden, "Of Bytes and Bulletin Boards," *New York Times*, August 4, 1985, sec. 6, 34.

64. D'Arcy Fallon, "Boys Allegedly Target Female Classmates in Computer Prank/Dirty Message Sent Across Country," *Colorado Springs Gazette Telegraph*, July 8, 1993, A1.

65. John Markoff, "Networks of Computers at Risk from Invaders," *New York Times*, December 3, 1988, sec. 1, 8.

66. Kevin Driscoll, "Thou Shalt Love Thy BBS: Distributed Experimentation in Community Moderation," in *Computer Network Histories: Hidden Streams from the Internet Past*, ed. Gabriele Balbi, Gianluigi Negro, and Paolo Bory (Zürich: Chronos Verlag, 2019), 15–34.

67. Jennings is an anarchist, "guerilla ISP" operator, the founder of Shred of Dignity (skateboarders' rights group), publisher of the queercore zine *Homocore*, first webmaster for *Wired* magazine, and collector of Cold War computer artifacts.

68. Fred Hapgood, "Artificial Intelligence," *Omni* 10, no. 6 (1988): 106; Randy Bush, "FidoNet: Technology, Tools, and History," *Communications of the ACM* 36, no. 8 (1993): 31–35, https://doi.org/10.1145/163381.163383.

69. Hapgood, "Artificial Intelligence"; Kevin Driscoll, "Social Media's Dial-Up Roots," *IEEE Spectrum* 53, no. 11 (2016): 54–60, https://doi.org/10.1109/MSPEC.2016.7607028.

70. Matthias Röhr, "Home Computer on the Line: The West German BBS Scene and the Change of Telecommunications in the 1980s," *Media in Action: Interdisciplinary Journal on Cooperative Media* 1 (2017): 115–129, https://doi.org/10.25969/mediarep/16239.

71. Röhr.

72. The Constitution was amended in 1994 so that the federal government must ensure adequate telecommunications but was not in charge of providing it. The broad category of "electronic eavesdropping" almost made its way into the Basic Law in the 1950s and 1960s but was rejected at the constitutional level.

73. James G. Carr, "Wiretapping in West Germany," *American Journal of Comparative Law* 29, no. 4 (1981): 641n223, https://doi.org/10.2307/839756.

74. Paul M. Schwartz, "German and US Telecommunications Privacy Law: Legal Regulation of Domestic Law Enforcement Surveillance," *Hastings Law Journal* 54 (2002): 751–804, https://doi.org/10.2139/SSRN.425521; "Comparative Study on Wiretapping and Electronic Surveillance Laws in Major Foreign Countries," Library of Congress, Law Library Report (1975).

75. Klass v. Germany, App. No. 5029/71, 28 Eur. Ct. H.R. (ser. A) (1978), available at https://hudoc.echr.coe.int/fre?i=001-57510.

76. Bundesgerichtshof [BGH] [Federal Court of Justice] December 15, 1970, Bundesverfassungsgericht [BVerfG] 30 (1), 1970 (Ger.). The case is often called "Monitoring Opinion" or the "Privacy of Communications Case."

77. Klass, App. No. 58243/00, 28 Eur. Ct. H.R. para. 50.

78. Later, to meet the demands of the European Union, the Telecommunications Act (Telekommunikationsgesetz, TKG) of 1996 was passed. It requires providers to keep "inventory information," which is only name, address, telephone number, and non-dynamic IP address. Inventory information must be kept in databases, maintained by the telecom providers, and law enforcement requests to access it are made to a special independent authority within the Federal Ministry of Economic Affairs. The authority can access the information directly but must keep a log of retrievals. These requests can also be made directly to the telecommunication providers, free of judicial review, but the database that makes it easiest to identify those on the network is notably distinct. See Schwartz, "German and US Telecommunications Privacy Law," 751; Carl B. Kress, "The 1996 Telekommunikationsgesetz and the Telecommunications Act of 1996: Toward More Competitive Markets in Telecommunications in Germany and the United States," *Federal Communications Law Journal* 49, no. 3 (1996): 551; "Comparative Study on Wiretapping and Electronic Surveillance Laws in Major Foreign Countries."

79. Kai Denker, "Heroes Yet Criminals of the German Computer Revolution," in *Hacking Europe*, ed. Gerard Alberts and Ruth Oldenziel (Cham, Switzerland: Springer, 2014), 167–187; Sebastian Kubitschko, "Chaos Computer Club: The Communicative Construction of Media Technologies and Infrastructures as a Political Category," in *Communicative Figurations*, ed. Andreas Hepp, Andreas Breiter, and Uwe Hasebrink (London: Palgrave Macmillan, 2018), 81–100.

80. Sigmund P. Martin, "Controlling Computer Crime in Germany," *Information and Communications Technology Law* 5, no. 1 (1996): 5–28.

81. Cliff Stoll, *The Cuckoo's Egg: Tracking a Spy through the Maze of Computer Espionage* (New York: Simon and Schuster, 2005), 184.

82. Urlich Wuermeling, "New Dimensions on Computer-Crime—Hacking for the KGB—a Report," *Computer Law & Security Report* 5, no. 4 (1989): 20–21; "International Review of Penal Law—Computer Crime and Other Crimes Against Information Technology," *Computer Law and Security Review: The International Journal of Technology Law and Practice* 10, no. 1 (1994): 34.

83. Carol C. McCall, "Computer Crime Statutes: Are They Bridging the Gap Between Law and Technology," *Criminal Justice Journal* 11 (1988): 203; Giancarlo Taddei Elmi, "The Law on Computer Crime in Italy," *Information & Communications Technology Law* 6, no. 3 (1997): 249–265. Nonetheless, the Hanover Hackers were convicted of working as foreign agents for the intelligence service of a foreign power under section 99 of the criminal code, an old espionage law. The case was decided by the Higher Regional Court Celle (Oberlandesgericht Celle, OLG Celle) on May 2, 1990 (docket no. 4 StE 1/89). Many thanks to law librarian Jenny Gesley at the Library of Congress for tracking down the specifics of the charges and related material. According to privacy expert Rebecca Herold, German authorities struggled to prosecute the case because "of unclear laws on computer crime and a lack of awareness about what crimes can be committed or aided by computer." Herold explains how the hackers printed out documents, making the scale of the damage uncertain. Rebecca Herold, *The Privacy Papers: Managing Technology, Consumer, Employee and Legislative Actions* (Boca Raton, FL: CRC Press, 2001).

84. Michael Dobbs, "Minitel Is New French Revolution," *Washington Post*, December 25, 1986.

85. D. Collingwood Nash and J. B. Smith, "Interactive Home Media and Privacy," Office of Policy Planning, Federal Trade Commission (January 1981), 59.

86. The executive was Peter Winter, president of Online International Inc., former executive editor of Keycom, and an architect of Ceefax, the British Broadcasting Corporation's online information system. Epstein, "Et voila! Le Minitel," 46.

87. Deanna C. Nash and David A. Bollier, "Protecting Privacy in the Age of Hometech," *Technology Review* 83, no. 8 (1981): 66–75; Alan F. Westin, "Home Information System: The Privacy Debate," *Datamation* 28, no. 7 (1982): 100–114.

88. Mindy Elisa Watchtel, "Videotex: A Welcome New Technology or an Orwellian Threat to Privacy," *Cardozo Arts & Entertainment Law Journal* 2 (1983): 290.

89. Mark Nollinger, "America, Online!," *Wired*, September 1, 1995, https://www.wired.com/1995/09/aol-2/.

90. Kevin Driscoll, *The Modem World: A Prehistory of Social Media* (New Haven, CT: Yale University Press, 2022), 183, 262; Bradley Fidler, "Eternal October and the End of Cyberspace," *IEEE Annals of the History of Computing* 39, no. 1 (2017): 6–7; Wendy M. Grossman, *Net.Wars* (New York: NYU Press, 1997), chapter 3.

91. Users were able to choose their own individual screen names, which had to be between 3 to 10 characters made up of letters, numbers, and/or spaces. Each screen name had an associated password of 4 to 8 letters and/or numbers that the user would use to log in to their account. Ruth Maran, *America Online Simplified* (Wiley, 2000), 162. Members of AOL could also input basic information on their user profile that anyone on the network could access. This information included categories such as name, location, birthdate, sex, marital status, hobbies, computers, occupation, and a personal quote. This profile information was optional. All categories—except sex, which contained radial buttons for male, female, or no response—were empty text boxes for users to fill. When users chose to fill out their profile information, it became publicly searchable within the AOL member directory. While searching for users based on categorical information via the member directory allowed users to locate the screen names of certain people, if a user already knew an individual's screen name, they could input it in "Locate AOL Member Online" to find them online. The tech-minded hobbyists who created online communities via bulletin boards were used to cultures of collaboration and openness, and they viewed this popular practice of identifying oneself via a screen name as an embrace of illicit activity. Grossman, *Net Wars*; DAwn, "The real reason kiddie porn dealers use AOL [Online forum comment]," June 3, 1998, Message posted to https://groups.google.com/forum/#!searchin/alt.aol-sucks/anonymous%7Csort:date/alt.aol-sucks/u5i6rgHDfTE/YthT4vsewgwJ; C. Johnson, "How do I stop my friends from getting AOL? [Online forum comment]," February 3, 1998, Message posted to https://groups.google.com/forum/#!searchin/alt.aol-sucks/anonymous%7Csort:date/alt.aol-sucks/nGr5PD_3HZI/LUWwJEadqxIJ.

92. "AOL's 'Walled Garden,'" *Wall Street Journal*, September 4, 2000, https://www.wsj.com/articles/SB968104011203980910.

93. Banks, *On the Way to the Web*, 192.

94. Chris Reidy, "Computer Flap: Is Speech Free on Prodigy?," *Boston Globe*, January 30, 1991.

95. "Prodigy Moves to Allay User Concerns on Privacy," *Wall Street Journal*, July 30, 1991, B4.

96. "Big Brother Not Watching, Prodigy Says," *Austin American Statesman*, May 4, 1991, F1.

97. Orin S. Kerr, "The Mosaic Theory of the Fourth Amendment," *Michigan Law Review* 111 (2012): 311–354; David Gray and Danielle Keats Citron, "A Shattered Looking Glass: The Pitfalls and Potential of the Mosaic Theory of Fourth Amendment Privacy," *North Carolina Journal of Law & Technology* 14, no. 2 (2012): 381–430; Paul Ohm, "The Many Revolutions of Carpenter," *Harvard Journal of Law & Technology* 32 (2018): 357–416.

98. Schwartz, "German and US Telecommunications Privacy Law," 766.

99. Anuj C. Desai, "Wiretapping before the Wires: The Post Office and the Rebirth of Communications Privacy," *Stanford Law Review* 60 (2007): 553–592; Efrat Nechushtai, "Making Messages Private: The Formation of Postal Privacy and Its Relevance for Digital Surveillance," *Information & Culture* 54, no. 2 (2019): 133–158.

100. George B. Prescott, *History, Theory, and Practice of the Electric Telegraph* (Boston: Ticknor and Fields, 1860), 338, cited in Thomas Jepsen, "'A New Business in the World': The Telegraph, Privacy, and the U.S. Constitution in the Nineteenth Century," *Technology and Culture* 59, no. 1 (2018): 95–125, doi:10.1353/tech.2018.0007; Susan P. Crawford, "Transporting Communications," *Boston University Law Review* 89 (June 2009): 879.

101. Richard R. John, *Network Nation: Inventing American Telecommunications* (Cambridge, MA: Harvard University Press, 2010), 144.

102. Warren and Brandeis wrote, "Recent inventions and business methods call attention to the next step which must be taken for the protection of the person, and for securing to the individual what Judge Cooley calls the right 'to be let alone.'" Samuel D. Warren and Louis D. Brandeis, "The Right to Privacy," *Harvard Law Review* 4, no. 5 (1890): 15, https://doi.org/10.2307/1321160.

103. Olmstead v. United States 277 U.S. 438, 464 (1928).

104. Brian Hochman, *The Listeners: A History of Wiretapping in the United States* (Cambridge, MA: Harvard University Press, 2022); Colin Agur, "Negotiated Order: The Fourth Amendment, Telephone Surveillance, and Social Interactions, 1878–1968," *Information & Culture* 48, no. 4 (2013): 419–447.

105. Berger v. United States, 295 U.S. 78 (1935); Katz v. United States, 389 U.S. 347 (1967).

106. 90 H.R. 5037 Omnibus Crime Control and Safe Streets Act of 1968, § 2511(2)(c).

107. *Smith* case refers to telephone users; see quotes in Schwartz, "German and US Telecommunications Privacy Law." Indeed the 1968 act does refer to subscribers or users when defining devices that could be used to intercept communication as those provided to "the subscriber or user by communications common carrier for ordinary course of business."

108. Electronic Communications Privacy Act of 1986 (ECPA), 18 U.S.C. §§ 2510–2523.

109. *Oversight on Communications Privacy: Hearing before the Subcommittee on Patents, Copyrights, and Trademarks, Senate Committee on the Judiciary*, 98th Congress (September 12, 1984) (opening statement of Senator Patrick J. Leahy), 1.

110. *Electronic Communications Privacy Act: Hearing before the Senate Subcommittee on Patents, Copyrights, and Trademarks, Committee on the Judiciary*, 99th Congress (November 13, 1985) (testimony of Representative Carlos Moorhead), 37.

111. *Electronic Communications Privacy Act* (November 13, 1985) (statement of Philip M. Walker, vice chairman, Electronic Mail Association), 94.

112. *Electronic Communications Privacy Act* (November 13, 1985) (statement of P. Michael Nugent, government affairs counsel for Electronic Data Systems and representation of ADAPSO), 106.

113. *Electronic Communications Privacy Act: Hearings before the House Subcommittee on Courts, Civil Liberties, and the Administration of Justice, House Committee on the Judiciary*, 99th Congress (September 19, September 26, October 24, 1985; January 30, March 5, 1986) (letter from Leslie C. Seeman, General Counsel, The Source Information Network to Hon. Robert W. Kastenmeier), 409.

114. 18 U.S. Code § 2511(3)(b).

115. 18 U.S. Code § 2701(c)(2).

116. 18 U.S. Code § 2702(b)(3).

117. 18 U.S. Code § 2702(c)(6).

118. *Counterfeit Access Device and Computer Fraud and Abuse Act: Hearing before the House Subcommittee on Crime, Committee on the Judiciary*, 98th Congress (November 10, 1983) (statement of Peter C. Waal, vice president of marketing, GTE Telenet), 185; Report to accompany H.R. 4718, May 22, 1986, Report 99–612, 3.

119. 18 U.S. Code § 1030.

120. Florida Representative Bill Nelson stated in his testimony, "We have had this legislation in the Congress almost 4 years. But, interestingly, it has taken the 414s, the computer hackers from Milwaukee, WI, to focus national attention that there is a problem." *Counterfeit Access Device and Computer Fraud and Abuse Act: Hearing before the House Subcommittee on Crime, Committee on the Judiciary*, 98th Congress (September 29, 1983) (testimony of Congressman Bill Nelson), 25.

121. Alex Orlando, "The Story of the 414s: The Milwaukee Teenagers Who Became Hacking Pioneers," *Discover Magazine*, October 10, 2020, https://www.discovermagazine.com/technology/the-story-of-the-414s-the-milwaukee-teenagers-who-became-hacking-pioneers.

122. "Computer Capers: Trespassing in the Information Age—Pranks or Sabotage?," *Newsweek*, September 5, 1983, cover.

123. *Computer and Communications Security and Privacy: Hearing before the Subcommittee on Transportation, Aviation, and Materials, House Committee on Science and Technology*, 98th Congress (September 26, 1983) (statement of Neal Patrick, Rufus King High School student), 14–16. Joy Lisi Rankin describes the many education networks built in the Midwest. Joy Rankin, "From the Mainframes to the Masses: A Participatory Computing Movement in Minnesota Education," *Information & Culture* 50, no. 2 (2015): 197–216, https://doi.org/10.1353/lac.2015.0009; for a critique of "the user" in Dartmouth's call to "respect the user" on their network, see Joy Lisi Rankin, *A People's History of Computing in the United States* (Cambridge, MA: Harvard University Press, 2018).

124. *The 414s: The Original Teenage Hackers*, directed by Michael T. Vollmann (2015).

125. *The Computer Fraud and Abuse Act: Hearings before the Senate Judiciary Committee*, 99th Congress (April 16, 1986) (testimony of Joseph Tompkins, chair of the American Bar Association Task Force on Computer Crime), 39.

126. *Counterfeit Access Device and Computer Fraud and Abuse Act: Hearing before the Subcommittee on Crime, House Judiciary Committee*, 98th Congress (March 28, 1984) (testimony of George Minot, vice president of CompuServe), 295, 320–321. Note that George Minot, vice president of CompuServe, was representing the Videotex Industry Association, which was only two years old at the time and represented 120 companies trying to popularize Videotex services.

127. Banks, *On the Way to the Web*.

128. Mitch Ratcliffe, "Euronalysis: We're Over Here, It's Over There," *Digital Media* 4, no. 12 (May 10, 1995).

129. "Europe's Online Services Prepare for Advertising," *Euromarketing* 9, no. 14 (December 12, 1995); "Europeans Find U.S. Multimedia Industry Threatening," *Multimedia Week* 4, no. 3 (January 1995).

130. Niels Kerssens, "Rethinking Legacies in Internet History: Euronet, Lost (Inter) Networks, EU Politics," *Internet Histories* 4, no. 1 (2020): 32–48.

131. "Ask for Diane," *Economist*, February 16, 1980, 57.

132. "Ask for Diane."

133. "Ask for Diane," 57.

134. Mark Wheeler, "Supranational Regulation: Television and the European Union," *European Journal of Communication* 19, no. 3 (2004): 349–369.

135. Viktor Mayer-Schonberger and Mathias Strasser, "Closer Look at Telecom Deregulation: The European Advantage," *Harvard Journal of Law & Technology* 12 (1998): 561–588.

136. Recommendation No. R(95) 4 of the Committee of Ministers of the Council of Europe to Member States on the Protection of Personal Data in the Area of Telecommunication Services, with particular reference to Telephone Services.

137. Procedure 2000/0189/COD COM (2000) 385: Proposal for a Directive of the European Parliament and of the Council concerning the processing of personal data and the protection of privacy in the electronic communications sector.

138. ePrivacy Directive, Directive 2002/58/EC of the European Parliament and of the Council of 12 July 2002 concerning the processing of personal data and the protection of privacy in the electronic communications sector (Directive on Privacy and Electronic Communications, ePrivacy Directive) OJ L 201, 31.7.2002, Article 5(3), (3).

139. ePrivacy Directive, Article 2(a).

140. From the European Informatics Network in 1976, European Unix Network in 1982, and European Academic and Research Network in 1983 to the Euronet of 1984 (when its technical standards were distributed) and Ebone backbone of 1992, mix and match efforts of researchers, computer scientists, policymakers, and telecommunications stakeholders attempted to create a European internet; see, for example, the UK's relationship with the US's ARPANET. Kirstein, "Early Experiences with the ARPANET and Internet in the United Kingdom." Shahin, "A European History of the Internet."

141. Kerssens distinguishes legacy approaches to internet history that focus on US research projects, counterculture, computer markets, and information superhighway rhetoric from the discontinuities of other networks that fill in missing narratives and include complementary, supportive, and rival efforts. Kerssens, "Rethinking Legacies in Internet History," 32–48.

CHAPTER 4

1. Charles Petrie, "Robert Cailliau on the WWW Proposal: 'How It Really Happened,'" *Internet Computing Online* 2 (1997), https://web.archive.org/web/20110106041256/http://www.computer.org/portal/web/computingnow/ic-cailliau.

2. Petrie.

3. "History of the Web," World Wide Web Foundation, accessed September 29, 2023, https://webfoundation.org/about/vision/history-of-the-web/.

4. For a detailed history of the development of the web at CERN see James Gillies and Robert Cailliau, *How the Web Was Born: The Story of the World Wide Web* (New York: Oxford University Press, 2000) and Tim Berners-Lee, *Weaving the Web: The Original Design and Ultimate Destiny of the World Wide Web by Its Inventor* (New York: Harper, 1999).

5. Cian O'Luanaigh, "Dream Team of Web Developers to Recreate Line-Mode Browser," CERN, September 19, 2013, https://home.cern/news/news/computing/dream-team-web-developers-recreate-line-mode-browser.

6. Tim Berners-Lee, "Re: Qualifiers on Hypertext Links," alt.hypertext newsgroup, August 6, 1991, https://www.w3.org/People/Berners-Lee/1991/08/art-6484.txt.

7. Tim Smith and François Flückiger, "Licensing the Web," CERN, accessed August 24, 2023, https://home.cern/science/computing/birth-web/licensing-web.

8. Petrie, "Robert Cailliau on the WWW Proposal."

9. Christopher Leslie, "As We Could Have Thought: Deploying Historical Narratives of the Memex in Support of Innovation," *Technology and Culture* 61, no. 2 (2020): 480–511, https://doi.org/10.1353/tech.2020.0050.

10. Belinda Barnet, "Engelbart's Theory of Technical Evolution," *Continuum* 20, no. 4 (2006): 509–521, https://doi.org/10.1080/10304310600988302.

11. Theodor H. Nelson, "Complex Information Processing: A File Structure for the Complex, the Changing and the Indeterminate," *Proceedings of the 1965 20th National Conference* (1965): 84–100.

12. Gary Wolf, "The Curse of Xanadu," *Wired*, June 1, 1995, https://www.wired.com/1995/06/xanadu/.

13. Belinda Barnet, "Crafting the User-Centered Document Interface: The Hypertext Editing System (HES) and the File Retrieval and Editing System (FRESS)," *Digital Humanities Quarterly* 4, no. 1 (2010): 3.

14. Belinda Barnet, *Memory Machines: The Evolution of Hypertext* (London: Anthem Press, 2013). See also George P. Landow, *Hypertext 3.0: Critical Theory and New Media in an Era of Globalization* (Baltimore: Hopkins Press, 2006); James M. Nyce and Paul Kahn, eds., *From Memex to Hypertext: Vannevar Bush and the Mind's Machine* (San Diego: Academic Press, 1991).

15. Barnet, *Memory Machines*.

16. Barnet.

17. Barnet, 98.

18. Douglas R. Dechow and Daniele C. Struppa, eds., *Intertwingled: The Work and Influence of Ted Nelson* (Cham, Switzerland: Springer Nature, 2015).

19. Claire L. Evans, *Broad Band: The Untold Story of the Women Who Made the Internet* (Edmonton: Portfolio, 2018).

20. *BBC Master AIV Guide*, Acorn Computers Limited, 1986, https://www.domesday86.com/wp-content/uploads/2017/01/BBC-Master-AIV-User-Guide-1.pdf.

21. Wendy Hall, "Making Links: Everything Really Is Deeply Intertwingled," in *Intertwingled*, 75–82.

22. Evans, *Broad Band*.

23. Hall, "Making Links."

24. Tim Berners-Lee, "Talks," *W3C*, https://www.w3.org/People/Berners-Lee/Talks.html#1994.

25. Scott Carlson, "How Gopher Nearly Won the Internet," *Chronicle of Higher Education*, September 5, 2016, https://www.chronicle.com/article/how-gopher-nearly-won-the-internet/.

26. Marc Weber, "'Woodstock of the Web' at 25," *Computer History Museum*, May 24, 2019, https://computerhistory.org/blog/woodstock-of-the-web-at-25/; Michael Grobe, "An Early History of Lynx: Multidimensional Collaboration," April 10, 1997, http://people.cc.ku.edu/~grobe/early-lynx.html.

27. Grobe, "An Early History of Lynx."

28. Pei Y. Wei, "ViolaWWW Hypertext Browser," *W3C*, March 1992, https://www.w3.org/History/19921103-hypertext/hypertext/WWW/Viola/violaWWWAbout.html; Tim Berners-Lee, "Frequently Asked Questions," *W3C*, https://www.w3.org/People/Berners-Lee/FAQ.html#browser; Berners-Lee, *Weaving the Web*.

29. Accounts vary of how work on a browser came about, because of later legal claims about intellectual property and ownership.

30. Lisa Wainwright, Anne Balsamo, and Judy Malloy, "Colleen Bushell," in *New Media Futures: The Rise of Women in the Digital Arts*, ed. Donna J. Cox, Ellen Sandor, and Janine Fron (Champaign: University of Illinois Press, 2018), 93–98.

31. There's no shortage of histories of Mosaic. For an insider account, see Jim Clark and Owen Edwards, *Netscape Time: The Making of the Billion-Dollar Start-Up That Took On Microsoft* (New York: St. Martin's Press, 2000); for a historical perspective, see Brian McCullough, *How the Internet Happened: From Netscape to the iPhone* (New York: Liveright Publishing, 2018); "A Short History of the Web," CERN, accessed August 24, 2023, https://home.cern/science/computing/birth-web/short-history-web.

32. Petrie, "Robert Cailliau on the WWW Proposal."

33. Marc Andreessen, "Proposed New Tag: IMG," www-talk electronic mailing list, February 25, 1993, http://1997.webhistory.org/www.lists/www-talk.1993q1/0182.html.

34. Tim Berners-Lee, "Re: Proposed New Tag: IMG," www-talk electronic mailing list, February 26, 1993, 2:04 p.m., http://1997.webhistory.org/www.lists/www-talk.1993q1/0186.html; Tim Berners-Lee, "Re: Proposed New Tag: IMG," www-talk electronic mailing list, February 26, 1993, 6:12 p.m., http://1997.webhistory.org/www.lists/www-talk.1993q1/0191.html.

35. Colleen Bushell, "Illinois Storyteller Colleen Bushell," *Storied: University of Illinois*, August 20, 2020, https://storied.illinois.edu/center-of-the-universe/?fbclid=IwAR28DbFfZ1iOv2K63ynl3J3cw1d8RaO_fmTFdxV_lJtQjO-8LMQewVwslJY; Colleen Bushell, in discussion with the author, October 14, 2022.

NOTES

36. Megan Sapnar Ankerson, "How *Coolness* Defined the World Wide Web of the 1990s," *The Atlantic*, July 15, 2014, https://www.theatlantic.com/technology/archive/2014/07/how-coolness-defined-the-world-wide-web-of-the-1990s/374443/.

37. James Coates, "Prodigy Weaves Easy Access to World Wide Web," *Chicago Tribune*, February 11, 1995; Eric Gwinn, "CompuServe Finally Releases Internet-Access Update," *Chicago Tribune*, November 11, 1996; Daniel Akst, "A First Look at CompuServe's 3.0 Interface Lift—and a Mixed Opinion," *LA Times*, August 12, 1996, https://www.latimes.com/archives/la-xpm-1996-08-12-fi-33484-story.html; America Online, "America Online, Inc. Expands Internet Support," *PRNewswire*, June 2, 1994.

38. Matt Blitz, "Later, Navigator: How Netscape Won and Then Lost the World Wide Web," *Popular Mechanics*, April 4, 2019, https://www.popularmechanics.com/culture/web/a27033147/netscape-navigator-history/.

39. Jim Boulton, "9 Ideas That Shaped the Web," *Fast Company*, August 6, 2014, https://www.fastcompany.com/3034030/9-design-ideas-that-forever-changed-the-web.

40. Brian McCollough, "Netscape's Rosanne Siino," *Internet History Podcast* (interview audio), June 24, 2018, http://www.internethistorypodcast.com/2018/06/173-netscapes-rosanne-siino/.

41. McCollough.

42. "The Golden Geeks," *Time* (cover), February 19, 1996; "Lou Montulli: Sexiest Internet Mogul," *People*, November 15, 1999. For a discussion of magazine publications' role in shaping the image of Silicon Valley, see Delia Dumitrica and Georgia Gaden Jones, "Developing the 'Control Imaginary': *Time Magazine*'s Symbolic Construction of Digital Technologies," *International Journal of Communication* 14 (2020): 2519–2542.

43. Laurence Zuckerman, "With Internet Cachet, Not Profit, a New Stock Is Wall St.'s Darling," *New York Times*, August 10, 1995, https://www.nytimes.com/1995/08/10/us/with-internet-cachet-not-profit-a-new-stock-is-wall-st-s-darling.html.

44. Petrie, "Robert Cailliau on the WWW Proposal."

45. John Schwartz, "Giving Web a Memory Cost Its Users Privacy," *New York Times*, September 4, 2001, https://www.nytimes.com/2001/09/04/business/giving-web-a-memory-cost-its-users-privacy.html; Lou Montulli, "The Reasoning Behind Web Cookies," *The Irregular Musings of Lou Montulli*, May 14, 2013, https://montulli blogspot.com/2013/05/the-reasoning-behind-web-cookies.html.

46. HotWired.com was launched in 1994 and Organic was one of the first businesses building commercial websites, cofounded in 1993 with Jonathan Nelson, Cliff Skolnick, and Matthew Nelson. ("Everyone had a side hustle in those days.") Brian Behlendorf, in discussion with the author, September 18, 2019. Behlendorf created an electronic mailing list to contribute to improving the NCSA's HTTP server and worked closely with eight other programmers. In early 1995, they branched

off to create the Apache HTTP server, which quickly became the web's most popular server and continues to be today. In 1999, the open-source project became the Apache Software Foundation, and Behlendorf served as its president. Behlendorf was later the chief technology officer of the World Economic Forum and has served on the board of the Mozilla Foundation and Electronic Frontier Foundation.

47. Brian Behlendorf, "Session Tracking," www-talk electronic mailing list, April 17, 1995, https://lists.w3.org/Archives/Public/www-talk/1995MarApr/0456.html.

48. Behlendorf.

49. Lou Montulli, "Re: Session Tracking," www-talk electronic mailing list, April 18, 1995, https://lists.w3.org/Archives/Public/www-talk/1995MarApr/0462.html.

50. The memory system was brought to fruition with help from John Giannandrea, Netscape's Scottish CTO who had also joined Andreessen in California to work on Netscape Mosaic. Giannandrea is currently Apple's Senior Vice President of Machine Learning and AI Strategy.

51. Brian Behlendorf, "Re: Session Tracking," www-talk electronic mailing list, April 18, 1995, https://lists.w3.org/Archives/Public/www-talk/1995MarApr/0468.html.

52. Doug McIlroy, email to the author, March 9, 2021.

53. Bob Morris left Bell Labs in 1986 to work for the National Security Agency. His son, the author of the Morris Worm launched in 1988, was the first felony conviction under the Computer Fraud and Abuse Act; McIlroy, email to the author.

54. McIlroy.

55. *Unix Time Sharing System: Programmer's Manual*, research version, 8th ed., vol 1 (Murray Hill, NJ: AT&T Bell Laboratories, February 1985); in 1987, McIlroy also published a reader to help interpret the program manuals that included a glossary with the same definition: Doug McIlroy, *A Research Unix Reader: Annotated Excerpts from the Programmer's Manual 1971–1986* (Murray Hill, NJ: AT&T Bell Laboratories, February 1987), 93.

56. Valerie Quercia and Tim O'Reilly, *X Window System User's Guide* (Sebastopol, CA: O'Reilly, 1988), appendix A, "System Management."

57. "Magic Cookie" Encyclopedia, *PCMag.com*, https://www.pcmag.com/encyclopedia/term/magic-cookie.

58. Dave Kristol and Lou Montulli, "HTTP State Management Mechanism (RFC 2109)," Internet Engineering Task Force, February 1997, https://tools.ietf.org/html/rfc2109.

59. Brian Behlendorf, "Session-ID Redux," www-talk electronic mailing list, July 25, 1995, https://lists.w3.org/Archives/Public/www-talk/1995JulAug/0192.html.

60. Marc Hedlund, "Re: Session-ID Redux," www-talk electronic mailing list, July 26, 1995, https://lists.w3.org/Archives/Public/www-talk/1995JulAug/0215.html.

61. Koen Holtman, in discussion with the author, August 12, 2019.

62. Koen Holtman, "Session-ID and Privacy Mechanisms," www-talk electronic mailing list, July 22, 1995, http://1997.webhistory.org/www.lists/www-talk.1995q3/0158.html.

63. Robert Robbins, "Re: Session-ID and Privacy Mechanisms," www-talk electronic mailing list, July 22, 1995, http://1997.webhistory.org/www.lists/www-talk.1995q3/0157.html; July 26, 1995, http://1997.webhistory.org/www.lists/www-talk.1995q3/0207.html.

64. Brian Behlendorf, "Session-ID Redux," www-talk electronic mailing list, July 25, 1995, https://lists.w3.org/Archives/Public/www-talk/1995JulAug/0192.html.

65. Kevin O'Connor, interview by Brian McCullough, Internet History Podcast, June 23, 2014, http://www.internethistorypodcast.com/2014/06/co-founder-of-doubleclick-kevin-oconnor/.

66. Kevin O'Connor, interview by Brian McCullough.

67. Edward L. Nash, *The Direct Marketing Handbook*, 3rd ed. (New York: McGraw, 1992); Edward L. Nash, *Database Marketing* (New York: McGraw, 1993).

68. Keith J. Kelly, "Millard Heading for Internet: Publishing Exec Hired by DoubleClick to Boost Brand," *Advertising Age*, August 19, 1996, 32.

69. Abbey Klaassen, "Fewer Than Six Degrees of Wenda Harris Millard," *Advertising Age*, April 27, 2008, 10.

70. Lisa Napoli, "DoubleClick Buys NetGravity," *New York Times*, July 14, 1999, https://archive.nytimes.com/www.nytimes.com/library/tech/99/07/cyber/articles/14advertising.html.

71. Others include Match Logic (bought and shuttered by Excite); 24/7 Media (merged with Real Media); AdForce and AdKnowledge (bought by CMGI); Avenue A, Burst! Media (bought UK ad network OTP Media and sold to Blinkx, now RhythmOne); DoubleClick (bought by Google); and Engage (bought by CMGI). Real Media merged with 24/7 Media to become 24/7 Real Media in 2001 and was acquired by the British advertising firm WPP in 2007.

72. The Pennsylvania Newspaper Association and the Mid-Atlantic Newspaper Services, Inc.

73. Engage became a subsidiary of CMGI (which in the 1980s sold mailing lists of university faculty and librarians to educational and professional publishers) and together the companies acquired AdKnowledge, Flycast, and AdForce to complement CMGI's growing internet portfolio and incubator projects.

74. Gil Beyda, in discussion with the author, September 24, 2019.

75. Michael W. Miller, "Firms Peddle Information from Driver's Licenses," *Wall Street Journal*, November 25, 1991, B1.

76. John Markoff, "More Threats to Privacy Seen as Computer Links Broaden," *New York Times*, June 1, 1988, A1.

77. Caroline Jack, "Circulating Database Marketing in the 1990s U.S. Business Press" (2019 ICA Communication History Division preconference paper, under review, on file with author); Laura J. Gurak, *Persuasion and Privacy in Cyberspace: The Online Protests over Lotus MarketPlace and the Clipper Chip* (New Haven, CT: Yale University Press, 1999).

78. Evan I. Schwartz, "Advertising Webonomics 101," *Wired*, February 1, 1996, https://www.wired.com/1996/02/webonomics-2/.

79. Peter H. Lewis, "An Ad (Gasp!) in Cyberspace," *New York Times*, April 19, 1994, D1.

80. Brian Behlendorf, in discussion with the author, September 18, 2019.

81. boyd would go on to become a powerful voice in contemporary privacy debates, as described in the next chapter; Mark Amerika and danah boyd, in discussion with the author, November 2020.

82. For additional coverage, see Grammatron's about page at https://www.grammatron.com/about.html.

83. Megan Sapnar Ankerson, *Dot-Com Design: The Rise of a Usable, Social, Commercial Web* (New York: NYU Press, 2018).

84. Ankerson, "How *Coolness* Defined the World Wide Web of the 1990s."

85. The screenshot in Ankerson's *Atlantic* article shows a cool sponsor of the day, but archived pages of the site at The Internet Archive show various forms of advertising over the years. Ankerson, "How *Coolness* Defined the World Wide Web of the 1990s."

86. Martin Nisenholtz, "How to Market on the 'Net: Simple Rules of the Road Will Help Advertisers Think Before They Leap," *Advertising Age*, July 11, 1994, 28.

87. Martin Nisenholtz, in discussion with the author, September 19, 2019.

88. Nisenholtz, "How to Market on the 'Net," 28. Privacy policies on the web have no definitive origin, nor do terms of service. Although copyright policies are found on many sites preserved from the 1990s, the oldest privacy policy, as we think of them today, found by digging in the Internet Archive was on the Warner Bros. website between April 6 and April 15, 1997. Between those dates, a "Legal/Privacy Information About This Site" was added to the site. The privacy portion states, "As a general policy, no personal information is automatically collected from visitors to its sites nor is so-called 'cookie' technology used by the sites created by WB Online." AOL's first privacy policy appeared between fall 1997 and April 25, 1998. Netscape and Wired posted privacy policies between spring and December 1998, and Yahoo! between spring and June 1998.

89. R. Tomlinson, "Internet Free Europe," *Fortune*, September 6, 1999, 165–172.

90. Diane Seo, "DoubleClick Sees Commercial Potential on Web," *LA Times*, March 26, 1998, https://www.latimes.com/archives/la-xpm-1998-mar-26-fi-32815-story.html.

91. There were exceptions. For instance, in the UK, Jamie Riddell began pushing agency clients to advertise on the web in 1996, starting a number of online advertising departments before opening his own digital advertising agency based on direct marketing principles called Cheeze with wife and managing director Katherine Jerman in 1999. Jaywing was another British startup that launched in 1999 from the direct marketing world by Martin Boddy and Andy Gardner. Both were acquired by the UK Digital Marketing Group (DMG) in 2007.

92. Christina Spurgeon, "Online Advertising," in *The Routledge Companion to Global Internet Histories*, ed. G. Goggin and M. McLelland (Abingdon, UK: Routledge, 2017), 387–398, 390.

93. Milton L. Mueller, *Networks and States: The Global Politics of Internet Governance* (Cambridge, MA: MIT Press, 2010).

94. Daniel Greene, *The Promise of Access: Technology, Inequality, and the Political Economy of Hope* (Cambridge, MA: MIT Press, 2021); notably, one of those leaders was SGI alum and Apple exec David Barram; Margaret O'Mara, *The Code: Silicon Valley and the Remaking of America* (New York: Penguin Books, 2020), 295.

95. O'Mara, *The Code*, 298.

96. O'Mara, 299 ("Placing the NII in Commerce's bailiwick signaled that the Information Superhighway was an economic policy with technology, and high-tech industries, right at the center.").

97. "The National Information Infrastructure: Agenda for Action," 58 Fed. Reg. 49,025 (September 21, 1993), available at https://clintonwhitehouse6.archives.gov/1993/09/1993-09-15-the-national-information-infrastructure-agenda-for-action.html; O'Mara, *The Code*, 299.

98. Improvement of Technical Management of Internet Names and Addresses; Proposed Rule, 63(34) Fed. Reg. 8825–8833, 8826 (February 20, 1998).

99. Improvement of Technical Management of Internet Names and Addresses.

100. Improvement of Technical Management of Internet Names and Addresses.

101. Improvement of Technical Management of Internet Names and Addresses.

102. Meghan Grosse, "Laying the Foundation for a Commercialized Internet: International Internet Governance in the 1990s," *Internet Histories* 4, no. 3 (2020): 271–286, https://doi.org/10.1080/24701475.2020.1769890.

103. Grosse, 276 (citing Domain Names (1998). Box 036, Folder 102015 13122 102015–004. Ira Magaziner Electronic Commerce Papers, William J. Clinton Presidential Library Archives, Little Rock, Arkansas, United States.).

104. Grosse, 276 (citing Domain Names (1998). Box 036, Folder 102015 13122 102015–004. Ira Magaziner Electronic Commerce Papers, William J. Clinton Presidential Library Archives, Little Rock, Arkansas, United States.).

105. Grosse, 271–286, 276 (citing an interview with Ira Magaziner given while in Japan. Japan (1997). Box 045, Folder 102024 13125 102024–002. Ira Magaziner Electronic Commerce Papers, William J. Clinton Presidential Library Archives, Little Rock, Arkansas, United States.).

106. Grosse, 271–286. Paul Starr calls activity in the DOC—including the crafting of legal and normative rules, and the specific design of communication, network, organizations, and institutions related to the media—the "constitutive moment" made by the Clinton administration. Paul Starr, *The Creation of the Media: Political Origins of Modern Communications* (New York: Basic Books, 2004), 4.

107. Peter N. Stearns, *Consumerism in World History: The Global Transformation of Desire*, 1st ed. (Abingdon, UK: Routledge, 2001), ix.

108. Stearns, ix.

109. Lizabeth Cohen, *A Consumers' Republic: The Politics of Mass Consumption in Postwar America* (New York: Knopf, 2003).

110. Cohen. See also Meg Jacobs, *Pocketbook Politics: Economic Citizenship in Twentieth-Century America* (Princeton, NJ: Princeton University Press, 2005).

111. Inger L. Stole, "'Selling' Europe on Free Enterprise: Advertising, Business and the US State Department in the Late 1940s," *Journal of Historical Research in Marketing* 8, no. 1 (2016): 44–64.

112. Sheryl Kroen, "A Political History of the Consumer," *Historical Journal* 47, no. 3 (2004): 731, https://doi.org/10.1017/S0018246X04003929.

113. Stole, "'Selling' Europe on Free Enterprise," 44–64.

114. Stole, 54–58.

115. Stole, 58.

116. Kroen, "A Political History of the Consumer," 732–734.

117. Kroen, 735.

118. Frank Trentmann, "Beyond Consumerism: New Historical Perspectives on Consumption," *Journal of Contemporary History* 39, no. 3 (2004): 373–401, https://doi.org/10.1177/0022009404044446.

119. President John. F. Kennedy: Consumer Bill of Rights, 108 Cong. Rec. 4263 (March 15, 1962).

120. The full quote reads: "The march of technology—affecting, for example, the food we eat, the medicines we take, and the many appliances we use in our homes—has increased the difficulties of the consumer along with his opportunities; and it

has outmoded many of the old laws and regulations and made new legislation necessary. The typical supermarket before World War II stocked about 1,500 separate food items—an impressive figure by any standard. But today it carries over 6,000. Ninety percent of the prescriptions written today are for drugs that were unknown twenty years ago. Many of the new products used every day in the home are complex. The housewife is called upon to be an amateur electrician, mechanic, chemist, toxicologist, dietitian, and mathematician—but she is rarely furnished the information she needs to perform these tasks proficiently."

121. "Preliminary Programme of the European Economic Community for a Consumer Protection and Information Policy," OJ C 92, 31975Y0425(02) April 25, 1975: 2–16, https://eur-lex.europa.eu/legal-content/EN/TXT/?uri=CELEX%3A31975Y0425%2802%29&qid=1614313023565.

122. "Preliminary Programme of the European Economic Community for a Consumer Protection and Information Policy."

123. "Preliminary Programme of the European Economic Community for a Consumer Protection and Information Policy."

124. Four main directives have passed: the Product Liability Directive 1985, Unfair Terms in Consumer Contracts Directive 1993, Unfair Commercial Practices Directive 2005, and the Consumer Rights Directive 2011.

125. Chris Jay Hoofnagle, *Federal Trade Commission Privacy Law and Policy* (Cambridge, UK: Cambridge University Press, 2016).

126. Hoofnagle, xiv.

127. Hoofnagle, 145

128. Myron W. Watkins's surveys found almost 60 percent of claims in the first decade of the agency's existence, 70 percent in 1925, and 91 percent in 1932. Myron W. Watkins, "The Federal Trade Commission a Critical Survey," *Quarterly Journal of Economics* 40, no. 4 (1926): 561–585; Myron W. Watkins, "An Appraisal of the Work of the Federal Trade Commission," *Columbia Law Review* 32, no. 2 (1932): 272, https://doi.org/10.2307/1114746.

129. James Bishop Jr. and Henry W. Hubbard, *Let the Seller Beware* (Washington, DC: National Press, 1969).

130. Inger Stole and Chris Hoofnagle both discuss the book and its prominence in public and policy discourse. Inger L. Stole, *Advertising on Trial: Consumer Activism and Corporate Public Relations in the 1930s* (Champaign: University of Illinois Press, 2010); Hoofnagle, *Federal Trade Commission*, 33.

131. Arthur Kallet and F. J. Schlink, *100,000,000 Guinea Pigs: Dangers in Everyday Foods, Drugs, and Cosmetics* (New York: Vanguard Press, 1932).

132. Stole, *Advertising on Trial*.

133. Inger L. Stole, "Consumer Protection in Historical Perspective: The Five-Year Battle over Federal Regulation of Advertising, 1933 to 1938," *Mass Communication & Society* 3, no. 4 (2000): 351–372, 365 (citing Gallup G. (February 9, 1940). An analysis of the study of consumer agitation. Box 76, Folder 11. National Broadcasting Corporations papers. Wisconsin Center Historical Archives, State Historical Society of Wisconsin, Madison).

134. Hoofnagle, *Federal Trade Commission*, 155 (citing *In re Matter of Lester Rothschild, Trading as Gen-O-Pak Co.*, 49 F.T.C. 1673 (1952); Rothschild v. F.T.C., 200 F.2d 39 (7th Cir. 1952)).

135. Arthur R. Miller, "Computers, Data Banks and Individual Privacy: An Overview," *Columbia Human Rights Law Review* 4, no. 1 (1972): 7; Arthur R. Miller, "Personal Privacy in the Computer Age: The Challenge of a New Technology in an Information-Oriented Society," *Michigan Law Review* 67, no. 6 (1969): 1148.

136. *Commercial Credit Bureaus: Hearings before the House Subcommittee on Invasion of Privacy, Committee on Government Operations*, 90th Congress (1968), 30–47.

137. *Commercial Credit Bureaus*, 49–51.

138. *Commercial Credit Bureaus*, 49–51.

139. Josh Lauer, *Creditworthy: A History of Consumer Surveillance and Financial Identity in America* (New York: Columbia University Press, 2017), 25.

140. Lauer, 222.

141. Lauer, 226.

142. Lauer, 260–262.

143. *Data Protection, Computers, and Changing Information Practices: Hearing before the Subcommittee on Government Information, Justice, and Agriculture, Committee on Government Operations*, 101st Congress (May 16, 1990), 44.

144. Richard B. Kielbowicz, "Origins of the Junk-Mail Controversy: A Media Battle over Advertising and Postal Policy," *Journal of Policy History* 5, no. 2 (1993): 250, http://doi.org/10.1017/S0898030600006734.

145. The Direct Mail Advertising Association (DMAA) was established in 1917. Its name was later changed to Direct Mail Marketing Association (DMMA) and then Direct Marketing Association (DMA). Keeping the acronym, the name was changed again to the Data & Marketing Association and then acquired by the Association of National Advertisers (ANA).

146. *Mailing Lists: Hearing before the Committee on Post Office and Civil Service*, 91st Congress (July 22, 1970), 4.

147. A story eerily similar to the story unearthed by journalist Charles Duhigg in 2012, which was a *New York Times* magazine cover story and now so commonly referenced it's become a privacy trope. Duhigg recounts how Andrew Pole marched

into a Target with a fist full of coupons for maternity and baby products that had been sent to his teenage daughter. The manager of the store apologized but Pole returned the apology when he found out his daughter was in fact pregnant. Target's aggressive pursuit to hook parents early has led them to detect very early indicators of pregnancy. Charles Duhigg, "How Companies Learn Your Secrets," *New York Times*, February 16, 2012, https://www.nytimes.com/2012/02/19/magazine/shopping-habits.html?_r=1&hp=&pagewanted=all.

148. *Mailing Lists*, 57.

149. *Treasury, Post Office, and General Government Appropriations for 1972: Hearings before a Subcommittee of the Committee on Appropriations* 92nd Congress (April 28, 1971) (statement of the Direct Mail Advertising Association, Inc.), 772.

150. US Department of Health, Education, and Welfare, *"Records, Computers and the Rights of Citizens," Report of the Secretary's Advisory Committee on Automated Personal Data Systems (HEW Report)*, Washington, DC: July 1973, 41, 294–296.

151. Andrew N. Case, "'The Solid Gold Mailbox': Direct Mail and the Changing Nature of Buying and Selling in the Postwar United States," *History of Retailing and Consumption* 1, no. 1 (2015): 28–46, https://doi.org/10.1080/2373518X.2015.1012863; "Address and List Management, Data Processing and Software, and Document Management," *Smithsonian National Postal Museum*, https://postalmuseum.si.edu/exhibition/america's-mailing-industry-industry-segments/address-and-list-management-data-processing.

152. US Department of Health, Education, and Welfare, *Hew Report*, 295 (Daniel H. Lufkin, "Appendix H: Mailing Lists").

153. US Department of Health, Education, and Welfare, *Hew Report*, 295 (Lufkin, "Appendix H").

154. Privacy Protection Study Commission, *Public Hearing Regarding the Study of Mailing Lists*, Washington, DC, November 12, 1975, 2.

155. Privacy Protection Study Commission, *Personal Privacy in an Information Society* (Washington, DC: US Government Printing Office, 1977), 125–154.

156. Privacy Protection Study Commission, *Personal Privacy in an Information Society*, 143–144.

157. Privacy Protection Study Commission, *Personal Privacy in an Information Society*, 151–153.

158. *In re Metromedia, Inc.*, No. C-1864 (February 17, 1971).

159. *In re Metromedia, Inc.*, No. C-1864 (February 17, 1971).

160. Hoofnagle, *Federal Trade Commission*, 156 (Hoofnagle notes that when looking closely at the orders, companies could condition providing services to clients on their consent to information practices).

161. Hoofnagle, 31, 60–66.

162. Hoofnagle, 71.

163. Hoofnagle, 145.

164. Tim Jackson, "This Bug in Your PC Is a Smart Cookie," *Financial Times*, February 12, 1996, 15.

165. Jackson, 15.

166. David Whalen, in discussion with the author, February 18, 2019.

167. Lou Montulli, "Re: Session Tracking," www-talk electronic mailing list, April 20, 1995, https://lists.w3.org/Archives/Public/www-talk/1995MarApr/0503.html.

168. Dave Kristol, "HTTP Cookies: Standards, Privacy, and Politics," *ACM Transactions on Internet Technology* 1, no. 2 (November 2001): 151–198.

169. Dave Kristol and Lou Montulli, "Proposed HTTP State Management Mechanism (Internet Draft)," Internet Engineering Task Force, February 16, 1996, https://web.archive.org/web/19990429122709/http://portal.research.bell-labs.com/~dmk/cookie-2.3.txt; Dave Kristol and Lou Montulli, "Proposed HTTP State Management Mechanism (Internet Draft)," Internet Engineering Task Force, February 19, 1996, https://web.archive.org/web/19990203222345/http://portal.research.bell-labs.com/~dmk/cookie-2.4.txt.

170. Ted Hardie, "Re: Unverifiable Transactions / Cookie Draft," www-talk electronic mailing list, March 18, 1997, https://lists.w3.org/Archives/Public/ietf-http-wg-old/1997JanApr/0472.html. Hardie started his work at Stanford Research Institute's Network Information Center, was at the time working at NASA's NIC, and would later lead teams at Qualcomm, Panasonic, and Google and become a vice president at Cisco.

171. EPIC was founded by Marc Rotenberg in 1994 after establishing and directing the Computer Professionals for Social Responsibility in Washington DC.

172. Dwight Merriman, "Unverifiable Transactions / Cookie Draft," www-talk electronic mailing list, March 13, 1997, https://lists.w3.org/Archives/Public/ietf-http-wg-old/1997JanApr/0416.html.

173. EPIC, "Net Users Urge Standards Group to Protect Privacy," EPIC press release, April 7, 1997, https://epic.org/privacy/internet/cookies/ietf_letter.html.

174. Quoting ADSmart's Sue Doyle. Rick Bruner, "Interactive: 'Cookie' Proposal Could Hinder Online Advertising; Privacy Backers Push for More Data Controls," *Advertising Age*, March 31, 1997, http://adage.com/article/news/interactive-cookie-proposal-hinder-online-advertising-privacy-backers-push-data-controls/68730/.

175. Bruner, "Interactive."

176. Dave Kristol, "HTTP Cookies: Standards, Privacy, and Politics," *ACM Transactions on Internet Technology* 1, no. 2 (November 2001): 151–198.

NOTES 241

177. Keith Moore and Ned Freed, "Use of HTTP State Management (RFC 2964)," Internet Engineering Task Force, October 2000.

178. Montulli, "The Reasoning Behind Web Cookies"; Rick E. Bruner, "Advertisers Win One in Debate over 'Cookies': Netscape Move May Settle Sites' Concern over Controversial Targeting Tool," *Advertising Age* 62 (May 12, 1997).

179. Lynette Millett, Batya Friedman, and Edward Felten, "Cookies and Web Browser Design: Toward Realizing Informed Consent Online," *Proceedings of the SIGCHI Conference on Human Factors in Computing Systems* (2001): 46–52.

180. Montulli, "The Reasoning Behind Web Cookies."

181. Montulli.

182. Federal Trade Commission, *Privacy Online: A Report to Congress*, June 1998, https://www.ftc.gov/sites/default/files/documents/reports/privacy-online-report-congress/priv-23a.pdf.

183. Federal Trade Commission, *Privacy Online*, 7.

184. Federal Trade Commission, *Online Privacy*, July 13, 1999, https://www.ftc.gov/sites/default/files/documents/public_statements/prepared-statement-federal-trade-commission-online-privacy/pt071399.pdf.

185. Federal Trade Commission, *Online Privacy*, 3.

186. National Telecommunications and Information Administration (NTIA), "Elements of Effective Self-Regulation for Protection of Privacy" (discussion draft), January 27, 1998.

187. U.S. Department of Energy, Computer Incident Advisory Capability (CIAC), "Internet Cookies," information bulletin I-034, March 12, 1998.

188. Federal Trade Commission, "Re: DoubleClick, Inc." (letter explaining that the investigation concluded after finding no personally identifiable information other than for those purposes stated in the privacy policy), January 22, 2001, https://www.ftc.gov/sites/default/files/documents/closing_letters/doubleclick-inc./doubleclick.pdf.

189. Founding members were 24/7 Media, AdForce, AdKnowledge, Avenue A, Burst! Media, DoubleClick, Engage, and MatchLogic.

190. Jules Polonetsky (DoubleClick's first chief privacy officer), in discussion with the author, May 14, 2019; Marc Rotenberg (founder of EPIC), in discussion with the author, August 29, 2019; Network Advertising Initiative (NAI), "About the NAI," accessed August 25, 2023, https://thenai.org/about/.

191. Federal Trade Commission, "Federal Trade Commission Issues Report on Online Profiling," July 27, 2000, https://www.ftc.gov/news-events/press-releases/2000/07/federal-trade-commission-issues-report-online-profiling.

192. Bob Gellman and Pam Dixon, "Many Failures: A Brief History of Privacy Self-Regulation in the United States," *World Privacy Forum Report*, October 14, 2011, http://www.worldprivacyforum.org/wp-content/uploads/2011/10/WPFselfregulation history.pdf.

CHAPTER 5

1. William Quinn and John D. Turner, *Boom and Bust: A Global History of Financial Bubbles* (Cambridge, UK: Cambridge University Press, 2020).

2. One of America's most infamous attorneys of the twentieth century, Lerach helped take down Enron and later was sentenced to two years in prison and disbarred for concealing illegal payments to plaintiffs. Patrick Dillon and Carl M. Cannon, *Circle of Greed: The Spectacular Rise and Fall of the Lawyer Who Brought Corporate America to Its Knees* (New York: Crown, 2010). Melyvn Weiss also went to prison for misconduct at their firm and was disbarred. But these men were heroes to the plaintiffs they fought for—letters of support inundated the court when charges came. According to one judge, "Mel Weiss did not invent securities class actions, but he brought them to a prominence and impact unmatched by any other lawyer of his time." Sam Roberts, "Melvyn Weiss, Lawyer Who Fought Corporate Fraud, Dies at 82," *New York Times*, February 5, 2018, https://www.nytimes.com/2018/02/05/obituaries/melvyn-weiss-lawyer-who-fought-corporate-fraud-dies-at-82.html; James Freeman, "The Most Feared Man in Silicon Valley," *Forbes*, July 6, 2000, https://www.forbes.com/2000/07/06/freeman_0706.html#45d97c6a2fcb; Jeffrey Toobin, "The Man Chasing Enron," *New Yorker* 78, no. 26 (September 9, 2002): 86. Under the headline "Bloodsucking Scumbag," *Wired* explained in 1996 that Lerach "makes his living filing class-action lawsuits against high tech companies." Karen Donovan, "Bloodsucking Scumbag," *Wired*, November 1, 1996, https://www.wired.com/1996/11/es-larach/. The *New York Times* named him "the pit bull of Silicon Valley," stating flatly that Lerach was the "most hated man in high tech." Lawrence M. Fisher, "Profile: William S. Lerach: The Pit Bull of Silicon Valley," *New York Times*, September 19, 1993, nytimes.com/1993/09/19/business/profile-william-s-lerach-the-pit-bull-of-silicon-valley.html. One semiconductor company executive expressed, "I think Lerach and his ilk are a very low life form, somewhere below pond scum." Richard B. Schmitt, "Online Privacy: Alleged Abuses Shape New Law," *Wall Street Journal*, February 29, 2000, B1.

3. Schmitt, "Online Privacy."

4. *In re DoubleClick Inc. Privacy Litigation*, No.00-Civ-0641(NRB), Consolidated Amended Class Action Complaint 2000 WL 34403253 (S.D.N.Y.), May 26, 2000.

5. They also asserted common law claims of invasion of privacy, unjust enrichment, and trespass, as well as state consumer protection.

6. *In re DoubleClick Inc. Privacy Litigation*, 154 F.Supp.2d 497 (2001).

7. The court reasoned that DoubleClick itself was authorized as the user the communication was "intended for." Despite being stored on the human user's computer often without their knowledge, cookies were like business reply cards in magazines, according to the court. "These bar-codes and identification numbers are meaningless to consumers, but are valuable to companies in compiling data on consumer responses." The court does not extend this analogy to interrogate the bizarre scenario wherein a magazine subscription manager would come into a reader's home and take a business reply card out of a magazine sitting on a coffee table. This was assuming that cookies were "electronic communications in electronic storage," which the court does not. It argues electronic communication in electronic storage, according to the statute, requires it be held a) temporarily and b) by an electronic communication service.

8. *In re DoubleClick Inc. Privacy Litigation*, No.00-Civ-0641(NRB), 2000 WL 34403255 (S.D.N.Y.), May 26, 2000.

9. Fisher, "Profile: William S. Lerach."

10. "DoubleClick and Plaintiff Agree to Settle Class Action Privacy Litigation," *BusinessWire*, March 29, 2002.

11. *In re DoubleClick Inc. Privacy Litigation*, No.00-Civ-0641(NRB), Objections by Settlement Class Members Electronic Privacy Information Center, Junkbusters (S.D.N.Y.), May 2000.

12. Stephanie Miles, "DoubleClick Settles Investigation," *Wall Street Journal Europe*, August 28, 2002, A6.

13. *To Review the Federal Trade Commission's Survey of Privacy Policies Posted by Commercial Web Sites: Hearings before the Senate Committee on Commerce, Science, and Transportation*, 106th Congress (May 25, 2000); *Internet Privacy: Hearings before the Senate Committee on Commerce, Science, and Transportation*, 106th Congress (May 24, 2000), video available from C-SPAN: https://www.c-span.org/video/?157367-1/internet-privacy.

14. Federal Trade Commission, *Privacy Online: A Report to Congress*, 1998, https://www.ftc.gov/sites/default/files/documents/reports/privacy-online-report-congress/priv-23a.pdf; Federal Trade Commission, *Self-Regulation and Privacy Online: A Report to Congress*, 1999, https://www.ftc.gov/system/files/documents/reports/self-regulation-privacy-onlinea-federal-trade-commission-report-congress/1999self-regulationreport.pdf; Federal Trade Commission, *Privacy Online: Fair Information Practices in the Electronic Marketplace: A Report to Congress*, 2000, https://www.ftc.gov/sites/default/files/documents/reports/privacy-online-fair-information-practices-electronic-marketplace-federal-trade-commission-report/privacy2000.pdf.

15. *Internet Privacy: Hearings before the Senate Committee on Commerce, Science, and Transportation*, 106th Congress (May 24, 2000).

16. *Internet Privacy.*

17. *Internet Privacy.*

18. *Consumer Internet Privacy Enhancement Act*, S.2928, 106th Congress (1999–2000).

19. *Protection for Private Blocking and Screening of Offensive Material*, U.S. Code 47 (1996) § 230 (known as Section 230 of the 1996 Communication Decency Act, which was part of the 1996 Telecommunications Act).

20. *Internet Privacy: Hearings before the Senate Commerce, Science, and Transportation Committee*, 106th Congress (May 24, 2000).

21. "With This Law, You Can Spam," *Wired*, January 23, 2004, https://www.wired.com/2004/01/with-this-law-you-can-spam/; Jasmine E. McNealy, "Spam and the First Amendment Redux: Free Speech Issues in State Regulation of Unsolicited Email," *Communication Law & Policy* 22, no. 3 (2017): 351–373, https://doi.org/10.1080/10811680.2017.1331641.

22. Matthew Crain, *Profit Over Privacy: How Surveillance Advertising Conquered the Internet* (Minneapolis: University of Minnesota Press, 2021), 139 (citing Ken Auletta, *Googled: The End of the World as We Know It* (New York: Penguin, 2009), 91.

23. Crain, *Profit Over Privacy*, 144.

24. *An Examination of the Google-Doubleclick Merger and the Online Advertising Industry: What Are the Risks for Competition and Privacy? Hearings before the Subcommittee on Antitrust, Competition Policy, and Consumer Rights, Committee on the Judiciary*, 110th Congress (September 7, 2007) (statement and testimony of Marc Rotenberg), https://www.judiciary.senate.gov/imo/media/doc/Rotenberg%20Testimony%2009272007.pdf.

25. Jenny Lee, "The Google-DoubleClick Merger: Lessons from the Federal Trade Commission's Limitations on Protecting Privacy," *Communication Law and Policy* 25, no. 1 (2020): 77–103, https://doi.org/10.1080/10811680.2020.1690330.

26. Daniel J. Solove and Woodrow Hartzog, "The FTC and the New Common Law of Privacy," *Columbia Law Review* 114 (2014): 583–676.

27. Facebook, Inc., Docket No. C-4365, File No. 092-3184 (Fed. Trade Comm'n 2011) (agreement containing consent order), http://www.ftc.gov/os/caselist/0923184/111129Facebookagree.pdf.

28. Facebook, Inc., Docket No. C-4365, File No. 092-3184 (July 9, 2012) (decision and order), http://ftc.gov/os/caselist/0923184/120810facebookdo.pdf.

29. *In re Pharmatrak, Inc.*, 329 F. 3d 9 (1st Circ. 2003).

30. *In re Google Inc.*, 806 F. 3d 125, 132–133.

31. *In re Google Inc.*, 806 F. 3d, 143 (citing Caro v. Weintraub, 618 F.3d 94, 97 (2d Cir.2010) United States v. Pasha, 332 F.2d 193 (7th Cir.1964) ("[I]mpersonation of the intended receiver is not an interception within the meaning of the statute") and

NOTES   245

United States v. Pasha, 332 F.2d 193 (7th Cir.1964) ("[P]arty" would mean the person actually participating in the communication.)).

32. *In re Google Inc.*, 806 F. 3d.

33. Convention for the Protection of Human Rights and Fundamental Freedoms, November 4, 1950, 213 U.N.T.S. 222, available at https://www.echr.coe.int/Documents/Convention_ENG.pdf [hereafter ECHR], art. 8.

34. Charter of Fundamental Rights of the European Union (14 September 2000 CHARTER 4465/00 CONTRIB 319).

35. Charter of Fundamental Rights of the European Union, art. 7.

36. Charter of Fundamental Rights of the European Union, art. 8.

37. David Erdos, "Comparing Constitutional Privacy and Data Protection Rights within the EU" (University of Cambridge Faculty of Law Research Paper No. 21/2021), https://doi.org/10.2139/ssrn.3843653.

38. Directive 97/66/EC of the European Parliament and of the Council of 15 December 1997 concerning the processing of personal data and the protection of privacy in the telecommunications sector OJ L 24, 30.1.1998, Art. 1(2).

39. Directive 2002/21/EC of the European Parliament and of the Council of 7 March 2002 on a common regulatory framework for electronic communications networks and services (Framework Directive), Art 3.

40. Directive 98/34/EC of the European Parliament and of the Council of 22 June 1998 laying down a procedure for the provision of information in the field of technical standards and regulations. Directive 98/34/EC was repealed and replaced by Directive (EU) 2015/1535.

41. GDPR, Art. 2(a).

42. European Parliament (2000). First reading (co-decision procedure). Proposal for a European Parliament and Council directive concerning the processing of personal data and the protection of privacy in electronic communications sector (COM(2000) 385—C5–0439/2000—2000/ 0189(COD)), A5–0374/2001 [2001] OJ C140E/121 13.06.2002.

43. Article 29 Working Party (A29WP), *Recommendation 1/99 on Invisible and Automatic Processing of Personal Data on the Internet Performed by Software and Hardware*, 5093/98/EN/final WP 17 (February 23, 1999).

44. "Why the EU Should 'Save Our Cookie,'" *Precision Marketing* 11, November 16, 2001.

45. IAB, "Internet Advertising Bureau to Launch in Europe," IAB News press release, September 24, 1997, https://www.iab.com/news/internet-advertising-bureau-launch-europe/.

46. Sylvia Mercado Kierkegaard, "How the Cookies (Almost) Crumbled: Privacy & Lobbyism," *Computer Law & Security Review* 21, no. 4 (2005): 310–322.

47. Directive 2002/58/EC of the European Parliament and of the Council of 12 July 2002 concerning the processing of personal data and the protection of privacy in the electronic communications sector (Directive on Privacy and Electronic Communications, ePrivacy Directive) OJ L 201, 31.7.2002, Article 5(3); emphasis added.

48. Committee on Citizens' Freedoms and Rights, Justice and Home Affairs (Rapporteur: Marco Cappato), Recommendation for Second Reading on the Council common position for adopting a European Parliament and Council directive concerning the processing of personal data and the protection of privacy in the electronic communications sector (15396/2/2001—C500035/2002—2000/0189(COD)), A5-0130/2002, 22 April 2002, Amendment 4 Recital 25.

49. Kierkegaard, "How the Cookies (Almost) Crumbled," 320.

50. James Bainbridge, "Digital Cookies Allowed to Stay," *Media Week* 14, June 7, 2002.

51. Simon Roberts, "Users Are Still Wary of Cookies," *Computer Weekly* 24, June 20, 2002.

52. The same year AOL bought the German AdTech and Dave Morgan's later venture Tacoda, Yahoo! bought Right Media for $680 million, the British WPP bought 24/7 Real Media for $649 million, Microsoft bought aQuantive for $6 billion, and Google bought DoubleClick for $3.1 billion.

53. European Parliament, Legislative resolution of 24 September 2008 on the proposal for a directive of the European Parliament and of the Council amending Directive 2002/22/EC on universal service and users' rights relating to electronic communications networks, Directive 2002/58/EC concerning the processing of personal data and the protection of privacy in the electronic communications sector and Regulation (EC) No 2006/2004 on consumer protection cooperation (First Reading) (COM(2007) 0698—C6-0420/2007—2007/0248(COD)), 24.09.2008, Amendment 128; emphasis added.

54. European Parliament, Legislative resolution of 24 September 2008 on the proposal for a Directive of the European Parliament and of the Council amending Directive 2002/21/EC on a common regulatory framework for electronic communications networks and services, Directive 2002/19/EC on access to, and interconnection of, electronic communications networks and associated facilities, and Directive 2002/20/EC on the authorization of electronic communications networks and services (COM(2007)0697—C6-0427/2007—2007/ 0247(COD)) OJ C 8E, 14.1.2010; emphasis added.

55. European Council, Addendum to "I/A" note of 18 November 2009: Adoption of the proposal for a Directive of the European Parliament and of the Council amending the Directives 2002/21//EC on a common regulatory framework for electronic

communications networks and services, and 2002/20/EC on the authorisation of electronic communications networks and services (LA þ S) (third reading)—statements, 15864/09 ADD 1 REV 1.

56. Directive 2009/136/EC of the European Parliament and of the Council of 25 November 2009 amending Directive 2002/22/EC on universal service and users' rights relating to electronic communications networks and services, Directive 2002/58/EC concerning the processing of personal data and the protection of privacy in the electronic communications sector and Regulation (EC) No 2006/2004 on cooperation between national authorities responsible for the enforcement of consumer protection laws. OJ L 337, 18.12.2009.

57. The Article 29 Working Party was an advisory body composed of representatives from each of the Member States' data protection authorities, the European Data Protection Supervisor, and the European Commission. Under the GDPR, it was replaced by the European Data Protection Board in 2016; A29WP, *Opinion 3/2010 on the Principle of Accountability*, 00062/10/EN, WP 173 (July 13, 2010); A29WP, *Opinion 15/2011 on the Definition of Consent*, 01197/11/EN, WP 187 (July 13, 2011).

58. IAB, "Europe's Cookie Laws: E-Privacy Implementation Center," 2016, https://web.archive.org/web/20160617122010/http://www.iabeurope.eu/eucookielaws/.

59. Opinion 2/2010 on online behavioural advertising WP 171 (22.06.2010); Opinion 15/2011 on Consent WP 187 (13.07.2011); Opinion 04/2012 on Cookie Consent Exemption WP 194 (07.06.2012); Working Document 02/2013 providing guidance on obtaining consent for cookies WP 208 (02.10.2013).

60. For instance, the UK's Information Commissioner's Office did not require "prior" consent based on the change to 5(3) (UK Information Commissioner's Office, Guidance on the rules on use of cookies and similar technologies, May 6, 2012, https://ico.org.uk/media/for-organisations/documents/1545/cookies_guidance.pdf), but the Netherlands passed a strict law (Article 11.7(a)) in 2013 that deemed default privacy settings in the browser unacceptable forms of user consent—it was so unpopular that amendments were sought in 2014 and passed in March 2015. Wet van 4 februari 2015 tot wijziging van de Telecommunicatiewet (wijziging artikel 11.7a) (went into effect on March 10, 2015), art. 11.7a(1), https://zoek.officielebekendmakingen.nl/stb-2015-100.html. See also, Ronald Leenes and Eleni Kosta, "Taming the Cookie Monster with Dutch Law—a Tale of Regulatory Failure," *Computer Law & Security Review* 31 (2015): 317–335; Robert Bond, "The EU E-Privacy Directive and Consent to Cookies," *Business Lawyer* 68, no. 1 (2012): 215–224; and Joasia Luzak, "Much Ado About Cookies: The European Debate on the New Provisions of the ePrivacy Directive Regarding Cookies," *European Review of Private Law* 21 (2013): 221–245, https://doi.org/10.54648/erpl2013007.

61. Eduardo Ustaran, "Cookie Consent—What's Changed?," *International Association of Privacy Professionals (IAPP) Privacy Perspectives*, July 22, 2014, https://iapp.org/news/a/cookie-consent-whats-changed/.

62. Anita L. Allen, *Uneasy Access: Privacy for Women in a Free Society* (Lanham, MD: Rowman & Littlefield, 1988); Julie E. Cohen, "Examined Lives: Informational Privacy and the Subject as Object," *Stanford Law Review* 52 (1999): 1373–1438; Helen Nissenbaum, "Privacy as Contextual Integrity," *Washington Law Review* 79 (2004): 119–158.

63. Alessandro Acquisti, Laura Brandimarte, and George Loewenstein, "Privacy and Human Behavior in the Age of Information," *Science* 347, no. 6221 (2015): 509–514, https://doi.org/10.1126/science.aaa1465.

64. Joseph Turow, Michael Hennessy, and Nora Draper, "Persistent Misperceptions: Americans' Misplaced Confidence in Privacy Policies, 2003–2015," *Journal of Broadcasting & Electronic Media* 62, no. 3 (2018): 461–478.

65. danah boyd, *It's Complicated: The Social Lives of Networked Teens* (New Haven, CT: Yale University Press, 2014).

66. Cranor's impact on the field cannot be overstated. She directs institutes, labs, and graduate programs, served as the Chief Technologist at the FTC, and is a fellow of the ACM, IEEE, and AAAS. She is a founding figure in the usable privacy and security research community.

67. "In advance of the FTC Town Hall, 'Behavioral Advertising: Tracking, Targeting, and Technology,' to be held November 1–2, 2007 in Washington, D.C.," Letter to Secretary Donald S. Clark, Federal Trade Commission on behalf of Center for Democracy and Technology, Consumer Action, Consumer Federation of America, Electronic Frontier Foundation, Privacy Activism, Public Information Researcher, Privacy Journal, Privacy Rights Clearinghouse, and World Privacy Forum, http://www.worldprivacyforum.org/wp-content/uploads/2008/04/ConsumerProtections_FTC_ConsensusDoc_Final_s.pdf.

68. Chris Soghoian, "The History of the Do Not Track Header," Slight Paranoia, January 21, 2011, http://paranoia.dubfire.net/2011/01/history-of-do-not-track-header.html.

69. *Consumer Online Privacy: Hearing before the Senate Committee on Commerce, Science, and Transportation* (July 27, 2010) (testimony of Jay Rockefeller).

70. *Consumer Online Privacy*, 4 (testimony of Julius Genachowski).

71. Aleecia McDonald, "Stakeholders and High Stakes: Divergent Standards for Do Not Track," in *The Cambridge Handbook of Consumer Privacy*, ed. Evan Sellinger, Jules Polonestsky, and Omer Tene (Cambridge, UK: Cambridge University Press, 2018), 253.

72. World Wide Web Consortium, "W3C Workshop on Web Tracking and User Privacy: Workshop Report," March 2011, https://www.w3.org/2011/track-privacy/report.html.

73. McDonald, "Stakeholders and High Stakes," 258–259.

74. McDonald, 258–259.

75. Nicholas Doty, "Enacting Privacy in Internet Standards" (PhD diss., University of California, Berkeley, 2020), 113.

76. McDonald, "Stakeholders and High Stakes," 258.

77. McDonald, 263–264.

78. McDonald, 263–264.

79. McDonald, 263–264.

80. Chris Jay Hoofnagle, Jennifer M. Urban, and Su Li, "Privacy and Modern Advertising: Most US Internet Users Want 'Do Not Track' to Stop Collection of Data about Their Online Activities," Amsterdam Privacy Conference, 2012; Aleecia McDonald and Jon M. Peha, "Track Gap: Policy Implications of User Expectations for the 'Do Not Track' Internet Privacy Feature," *TPRC 2011* (September 25, 2011).

81. Robert Madelin, European Commission Letter to World Wide Web Consortium Tracking Protection Working Group, Ref. Ares(2012)743354, June 21, 2012, https://lists.w3.org/Archives/Public/public-tracking/2012Jun/att-0604/Letter_to_W3C_Tracking_Protection_Working_Group.210612.pdf.

82. Rigo Wenning to TPWG W3C electronic mailing list, "Consent," June 6, 2012, https://lists.w3.org/Archives/Public/public-tracking/2012Jun/0098.html.

83. Kimon Zorbas to TPWG W3C electronic mailing list, "Re: Examples of Successful Opt-In Implementations," June 14, 2012, https://lists.w3.org/Archives/Public/public-tracking/2012Jun/0413.html; Kimon Zorbas to TPWG W3C electronic mailing list, "Re: DNT Concerns," December 5, 2012, https://lists.w3.org/Archives/Public/public-tracking/2012Dec/0065.html#replies.

84. Rigo Wenning to TPWG W3C electronic mailing list, "Re: ACTION-174: Write Up Implication of Origin/* Exceptions in EU Context," June 6, 2012, https://lists.w3.org/Archives/Public/public-tracking/2012Jun/0109.html#replies.

85. S.B. 761, 112th Cong (2011).

86. A.B. 370, 113th Cong (2013).

87. Do Not Track Me Online Act of 2011, H.R. 654, 112th Cong. (2011); Commercial Privacy Bills of Rights Act of 2011, S. 799, 112th Cong. (2011); Consumer Privacy Protection Act of 2011, H.R. 1528, 112th Cong. (2011); Do-Not-Track Online Act of 2011, S. 913, 112th Cong. (2011); Do Not Track Kids Act, H.R. 1895, 112th Cong. (2011).

88. Meg Leta Jones and Jenny Lee, "Comparing Consent to Cookies: A Case for Protecting Non-Use," *Cornell International Law Journal* 53 (2020): 97–132.

89. plehegar, "WG closed," W3C/dnt GitHub, January 18, 2019, https://github.com/w3c/dnt/commit/5d85d6c3d116b5eb29fddc69352a77d87dfd2310.

90. "Turn 'Do Not Track' On or Off," Google Chrome Help, accessed September 28, 2023, https://support.google.com/chrome/answer/2790761?hl=en&co=GENIE.Platform%3DDesktop.

91. "U.S. Library of Congress, Congressional Research Service," *Origins and Impact of the Foreign Intelligence Surveillance Act (FISA) Provisions That Expired on March 15, 2020*, by Edward C. Liu, R40138 (March 31, 2021); National Security Agency, "National Security Agency: Missions, Authorities, Oversight and Partnerships," NSA press release no.: PA-026-18, August 9, 2013, https://web.archive.org/web/20210423180345/https://www.nsa.gov/news-features/press-room/Article/1618729/.

92. James Risen and Eric Lichtblau, "Bush Lets U.S. Spy on Callers Without Courts," *New York Times*, December 16, 2005, A1.

93. David Cole and Martin S. Lederman, "The National Security Agency's Domestic Spying Program: Framing the Debate," *Indiana Law Journal* 81 (2006): 1355–1425.

94. Interception and Disclosure of Wire, Oral, or Electronic Communications Prohibited, 18 U.S.C. § 2511(2)(a)(ii).

95. "NSA Multi-District Litigation," Electronic Frontier Foundation, accessed September 28, 2023, https://www.eff.org/cases/nsa-multi-district-litigation?page=5.

96. "Recent Legislation—Electronic Surveillance—Congress Grants Telecommunications Companies Retroactive Immunity from Civil Suits for Complying with NSA Terrorist Surveillance Program.—FISA Amendments Act of 2008, Pub. L. No. 110-261, 122 Stat. 2436," *Harvard Law Review* 122, no. 4 (February 2009): 1271–1278.

97. Pamela Hess, "UPDATE: US Intelligence Official: People Must Redefine Privacy," CNN, November 11, 2007, https://web.archive.org/web/20071117191002/https://money.cnn.com/news/newsfeeds/articles/djf500/200711112052DOWJONESDJONLINE000304_FORTUNE5.htm.

98. Letter from Michael B. Mukasey, Attorney General, and J. M. McConnell, Director of National Intelligence, to Nancy Pelosi, Speaker, US House of Representatives, June 19, 2008, https://web.archive.org/web/20090113202701/http://www.lifeandliberty.gov/docs/ag-dni-fisa-letter061908.pdf.

99. Amy Schatz, "Paul Camp, Liberals Unite on Spy Bill," *Wall Street Journal*, June 26, 2008, https://www.wsj.com/articles/SB121443403835305037?mod=googlenews_wsj.

100. "Senator Obama—Please Vote NO on Telecom Immunity—Get FISA Right," BarackObama.com, accessed August 24, 2023, https://web.archive.org/web/20080710013523/http://my.barackobama.com/page/group/SenatorObama-PleaseVoteAgainstFISA; Nick Juliano, "Group Urging FISA 'No' Vote Is Largest on Obama's Social Site," *The Raw Story*, July 3, 2008, https://web.archive.org/web/20080710135901/http://rawstory.com/news/2008/Group_urging_FISA_no_vote_largest_0703.html.

101. Roll Call Vote, On the Amendment (Dodd Amdt. No. 5064 to H.R. 6304 [Foreign Intelligence Surveillance Act of 1978]) 110th Congress, July 9, 2008, https://www.senate.gov/legislative/LIS/roll_call_lists/roll_call_vote_cfm.cfm?congress=110&session=2&vote=00164.

102. Maximilian Schrems, "Complaint against Facebook Ireland Ltd—23 'PRISM,'" to the Data Protection Commissioner of Ireland, June 25, 2013, https://noyb.eu/sites/default/files/2020-07/complaint-PRISM-facebook_2013.pdf.

103. Case C-362/14, Maximillian Schrems v. Data Protection Commissioner (October 6, 2015).

104. Daniel Castro, ITIF vice president, "US and EU Should Act Swiftly to Establish New Privacy Protections to Avoid Long-Term Digital Trade Disruption," Information Technology & Innovation Foundation press release, October 6, 2015, https://itif.org/publications/2015/10/06/us-and-eu-should-act-swiftly-establish-new-privacy-protections-avoid-long.

105. *Examining the EU Safe Harbor Decision and Impacts for Transatlantic Data Flows: Hearings before the Subcommittee on Commerce, Manufacturing, and Trade and the Subcommittee on Communications and Technology*, 114th Congress (November 3, 2015), https://energycommerce.house.gov/committee-activity/hearings/joint-hearing-on-examining-the-eu-safe-harbor-decision-and-impacts-for.

106. Daily Press Briefing by Press Secretary Josh Earnest, President Obama White House, October 6, 2015, https://obamawhitehouse.archives.gov/the-press-office/2015/10/06/daily-press-briefing-press-secretary-josh-earnest-1062015; *Examining the EU Safe Harbor Decision and Impacts for Transatlantic Data Flows*; Peter Swire, "Don't Strike Down the Safe Harbor Based on Inaccurate Views About U.S. Intelligence Law," *IAPP Privacy Perspectives*, October 5, 2015, https://iapp.org/news/a/dont-strike-down-the-safe-harbor-based-on-inaccurate-views-on-u-s-intelligence-law/; Joshua P. Meltzer, "The Court of Justice of the European Union in Schrems II: The Impact of GDPR on Data Flows and National Security," *Brookings*, August 5, 2020, https://www.brookings.edu/research/the-court-of-justice-of-the-european-union-in-schrems-ii-the-impact-of-gdpr-on-data-flows-and-national-security/#footref-6; Peter Sayer, "No Need to Panic: European Commission Upbeat about Safe Harbor Ruling," *Computer World*, October 6, 2015, https://www.computerworld.com/article/2989764/no-need-to-panic-european-commission-upbeat-about-safe-harbor-ruling.html.

107. Commission Implementing Decision (EU) 2016/1250 of 12 July 2016 pursuant to Directive 95/46/EC of the European Parliament and of the Council on the adequacy of the protection provided by the EU–U.S. Privacy Shield (notified under document C(2016) 4176), C/2016/4176, OJ L 207, 1.8.2016, http://eur-lex.europa.eu/legal-content/EN/TXT/?uri=uriserv%3AOJ.L_.2016.207.01.0001.01.ENG.

108. "Privacy Shield Overview," International Trade Administration (ITA), US Department of Commerce, https://www.privacyshield.gov/Program-Overview.

109. The White House, President Barack Obama, "Presidential Policy Directive—Signals Intelligence Activities," Policy Directive/PPD-28, January 17, 2014, https://obamawhitehouse.archives.gov/the-press-office/2014/01/17/presidential-policy-directive-signals-intelligence-activities; see also US Library of Congress, Congressional Research Service, *EU Data Transfer Requirements and U.S. Intelligence Laws: Understanding Schrems II and Its Impact on the EU–U.S. Privacy Shield*, by Chris D. Linebaugh and Edward C. Liu, R46724 (2017); The law has been criticized for offering nothing more than what US citizens get from the US Privacy Act, which isn't much when it comes to intelligence surveillance. David Bender, "The Judicial Redress Act: A Path to Nowhere," IAPP: The Privacy Advisor, December 17, 2015, https://iapp.org/news/a/the-judicial-redress-act-a-path-to-nowhere/.

110. European Parliament, Motion for a Resolution to wind up the debate on the statement by the Commission pursuant to Rule 123(2) of the Rules of Procedure on the adequacy of the protection afforded by the EU–US Privacy Shield (2016/3018(RSP)), March 3, 2017, https://www.europarl.europa.eu/doceo/document/B-8-2017-0235_EN.html.

111. The court was even less persuaded by the FCC's interest in promoting competition through restriction of CPNI. US West, Inc. v. FCC, 182 F.3d 1224 (10th Cir. 1999).

112. Federal Communications Commission, "Chairman Wheeler's Proposal to Give Broadband Consumers Increased Choice, Transparency & Security with Respect to Their Data," FCC press release, March 10, 2016, https://transition.fcc.gov/Daily_Releases/Daily_Business/2016/db0310/DOC-338159A1.pdf.

113. David L. Cohen, "FCC Action on Privacy Could Harm Consumers," *Comcast*, March 31, 2016, https://corporate.comcast.com/comcast-voices/fcc-action-on-privacy-could-harm-consumers.

114. Natasha Lomas, "FCC Proposes New Privacy Rules for ISPs," *TechCrunch*, April 4, 2016, https://techcrunch.com/2016/04/04/fcc-proposes-new-privacy-rules-for-isps/.

115. Federal Communications Commission, "Statement on Need for Comprehensive, Uniform Privacy Framework," FCC statement, February 24, 2017, https://www.fcc.gov/document/statement-need-comprehensive-uniform-privacy-framework; Federal Communications Commission, "FCC Moves to Ensure Consumers Have Uniform Online Privacy Protection," FCC press release, March 1, 2017, https://www.fcc.gov/document/fcc-moves-ensure-consumers-have-uniform-online-privacy-protection.

116. Margot E. Kaminski, "Binary Governance: Lessons from the GDPR's Approach to Algorithmic Accountability," *Southern California Law Review* 92 (2018): 1529–1616.

117. A29WP, *Guidelines on Consent under Regulation 2016/679*, 17/EN WP259 rev.01 (adopted on November 28, 2017, last revised and adopted on April 10, 2018), 17.

CHAPTER 6

1. Google explained that other browsers had started blocking third-party cookies by default, but "by undermining the business model of many ad-supported websites, blunt approaches to cookies encourage the use of opaque techniques . . . which can actually reduce user privacy and control." "Building a More Private Web: A Path Towards Making Third Party Cookies Obsolete," *Google Chromium Blog*, January 14, 2020, https://blog.chromium.org/2020/01/building-more-private-web-path-towards.html.

2. Worldwide: https://gs.statcounter.com/browser-market-share/all/worldwide/2020; in the US: https://gs.statcounter.com/browser-market-share/all/united-states-of-america/2020; in Europe: https://gs.statcounter.com/browser-market-share/all/europe/2020.

3. Shoshana Zuboff, *The Age of Surveillance Capitalism: The Fight for a Human Future at the New Frontier of Power* (New York: Public Affairs, 2019).

4. Klaus Wiedemann, "The ECJ's Decision in *'Planet49'* (Case C-673/17): A Cookie Monster or Much Ado About Nothing?," *International Review of Intellectual Property and Competition Law*, 51 (2020): 543–553, https://doi.org/10.1007/s40319-020-00927-w.

5. Case C-673/17, Bundesverband der Verbraucherzentralen und Verbraucherverbände—Verbraucherzentrale Bundesverband eV v. Planet49 GmbH, 2019 E.C.R.

6. The procedures for collecting cookies were based on article 82 of the French data protection law implementing article 5 (3) of the ePrivacy Directive.

7. CNIL, "Cookies and Trackers: What Does the Law Say?," CNIL press release, October 1, 2020, https://www.cnil.fr/fr/cookies-et-autres-traceurs/regles/cookies/que-dit-la-loi.

8. CNIL, "Cookies: The CNIL Fines Google a Total of 150 Million Euros and Facebook 60 Million Euros for Non-Compliance with French Legislation," CNIL press release, January 6, 2022, https://www.cnil.fr/en/cookies-cnil-fines-google-total-150-million-euros-and-facebook-60-million-euros-non-compliance.

9. Concerning: Complaint relating to IAB Transparency & Consent Framework, DOS-2019–01377, Gegevensbeschermingsautoriteit, Litigation Chamber, February 2, 2022, https://www.autoriteprotectiondonnees.be/publications/decision-quant-au-fond-n-21-2022-english.pdf#page111.

10. Cory Doctorow, "Consent Theater," *OneZero*, May 20, 2021, https://onezero.medium.com/consent-theater-a32b98cd8d96.

11. Biometric Information Privacy Act, 740 ILCS 14/.

12. Eliana Dockterman, "Prince Harry and Meghan, the Duke and Duchess of Sussex, Discuss Misinformation with Silicon Valley's Biggest Critics," *Time*, October 20, 2020, https://time.com/5901379/prince-harry-meghan-markle-tristan-harris-safiya-noble/.

13. "Black Future Month | March 20, 2019 Act 3 | Full Frontal on TBS," *Full Frontal with Samantha Bee* (video), March 21, 2019, https://youtu.be/AxpWvMrPqVs.

14. E. Barlow Keener, "Facial Recognition: A New Trend in State Regulation," *JD Supra*, May 2, 2022, https://www.jdsupra.com/legalnews/facial-recognition-a-new-trend-in-state-2773164/.

15. Senator Terry Link, Amendment to House Bill 6074, 09900HB6074sam001, Filed May 26, 2016, https://www.ilga.gov/legislation/99/HB/09900HB6074sam001.htm; Russell Brandom, "Someone's Trying to Gut America's Strongest Biometric Privacy Law," *The Verge*, May 27, 2016, https://www.theverge.com/2016/5/27/11794512/facial-recognition-law-illinois-facebook-google-snapchat; Megan Geuss, "Illinois Senator's Plan to Weaken Biometric Privacy Law Put on Hold," *Ars Technica*, May 27, 2016, https://arstechnica.com/tech-policy/2016/05/illinois-senators-plan-to-weaken-biometric-privacy-law-put-on-hold/.

16. *In re* Facebook Biometric Info. Privacy Litig., 185 F. Supp. 3d 1155 (N.D. Cal. 2016).

17. With an exception for US banks, which are given special treatment under BIPA.

18. Ryan Mac and Kashmir Hill, "Clearview AI Settles Suit and Agrees to Limit Sales of Facial Recognition Database," *New York Times*, May 9, 2022, https://www.nytimes.com/2022/05/09/technology/clearview-ai-suit.html.

19. Margot E. Kaminski, "The Case for Data Privacy Rights (Or 'Please, a Little Optimism')," *Notre Dame Law Review Online* (forthcoming 2022).

20. Ari Ezra Waldman, "Privacy, Practice, and Performance," *California Law Review* 110, no. 4 (2022), https://doi.org/10.15779/Z38JD4PQ3D.

21. Ari Ezra Waldman, "The New Privacy Law," *UC Davis Law Review Online* 55 (2021): 38.

22. Michael Veale (@mikarv), Twitter, March 8, 2021, 6:34 a.m., https://twitter.com/mikarv/status/1368887835189776384; Michael Veale (@mikarv), Twitter, March 8, 2021, 6:38 a.m., https://twitter.com/mikarv/status/1368888791033909248.

23. Darya Balybina, "What Is and What Isn't Subject to a DPIA Under GDPR? An Update," IAPP Privacy Advisor, February 14, 2020, https://iapp.org/news/a/what-is-and-what-isnt-subject-to-a-dpia-under-gdpr-an-update/.

24. Neil Richards and Woodrow Hartzog, "Taking Trust Seriously in Privacy Law," *Stanford Technology Law Review* 19 (2015): 431–472, https://doi.org/10.2139/ssrn.2655719; Jack M. Balkin, "Information Fiduciaries and the First Amendment," *UC Davis Law Review* 49, no. 4 (2016): 1183–1234; Neil Richards and Woodrow Hartzog, "A Duty of Loyalty for Privacy Law," *Washington University Law Review* 99 (2021): 961–1022.

25. Lina M. Khan and David E. Pozen, "A Skeptical View of Information Fiduciaries," *Harvard Law Review* 133, no. 2 (2020): 537.

26. Brian Fung, "The Unlikely Activist Behind the Nation's Toughest Privacy Law Isn't Done Yet," CNN, October 10, 2019, https://www.cnn.com/2019/10/10/tech/alastair-mactaggart/index.html.

27. Betsy Morris and Eliot Brown, "The Real-Estate Developer Who Took On the Tech Giants," *Wall Street Journal*, June 29, 2018, https://www.wsj.com/articles/the-real-estate-developer-who-took-on-the-tech-giants-1530308857.

28. Nicholas Confessore, "The Unlikely Activists Who Took On Silicon Valley—and Won," *New York Times Magazine*, August 14, 2018, https://www.nytimes.com/2018/08/14/magazine/facebook-google-privacy-data.html.

29. Cal. Civ. Code Section § 1798.135(a) ("Those that sell or share consumers' personal information or use or disclose consumers' sensitive personal information for purposes other than those authorized by subdivision (a) of Section 1798.121 . . .").

30. Text of Modified Regulations, California Consumer Privacy Act Regulations Proposed Text of Regulations, California Office of Attorney General, https://www.oag.ca.gov/sites/all/files/agweb/pdfs/privacy/ccpa-text-of-second-set-mod-031120.pdf?.

31. Gilad Edelman, "'Do Not Track' Is Back, and This Time It Might Work," *Wired*, October 7, 2020, https://www.wired.com/story/global-privacy-control-launches-do-not-track-is-back/; California Office of Attorney General, *Final Statement of Reasons*, February 3, 2023, 19–20, https://oag.ca.gov/sites/all/files/agweb/pdfs/privacy/ccpa-fsor.pdf.

32. Cal. Civ. Code Section § 1798.135(b)(2)(a).

33. Cal. Civ. Code Section § 1798.185(a)(19); Notably, some of the browser extensions and apps, like DuckDuckGo, already enable GPC by default. "Global Privacy Control (GPC) Enabled by Default in DuckDuckGo Apps & Extensions," DuckDuckGo News, January 28, 2021, https://spreadprivacy.com/global-privacy-control-enabled-by-default/.

34. Margot E. Kaminski, "Binary Governance: Lessons from the GDPR's Approach to Algorithmic Accountability," *Southern California Law Review* 92 (2018): 1529–1616.

35. Jack Nicas and Mike Isaac, "Facebook Takes the Gloves Off in Feud with Apple," *New York Times*, December 16, 2020 (updated April 26, 2021), https://www.nytimes.com/2020/12/16/technology/facebook-takes-the-gloves-off-in-feud-with-apple.html.

36. Luke Dormehl, "Full-Page Facebook Ad Accuses Apple of Changing the Internet for the Worse [Updated]," *Cult of Mac*, December 17, 2020, https://www.cultofmac.com/730598/full-page-facebook-ad-accuses-apple-of-changing-the-internet-for-the-worse/.

37. Richard H. Thaler and Cass R. Sunstein, *Nudge: Improving Decisions About Health, Wealth, and Happiness* (New Haven, CT: Yale University Press, 2008), 178–183.

38. Jill Lepore writes, "The machine, crammed with microscopic data about voters and issues could act as a macroscope. You could ask it any question about the kind of move that a candidate might make, and it would be able to tell you how voters, down to the tiniest segment of the electorate, would respond." Jill Lepore, *If Then: How the Simulmatics Corporation Invented the Future* (New York: Liveright Publishing, 2020), 92. In an interview with the Harvard Data Science Initiative, she clarifies, "A lot of it is just boosterism and flim-flam. What even is a 'People Machine'? It's just a program written in FORTRAN, and a bunch of data. The 'People Machine'—that's just boosterism." Francine Berman and Jill Lepore, "The People Machine: The Earliest Machine Learning? An Interview with Jill Lepore and Francine Berman," *Harvard Data Science Review*, January 29, 2021, https://doi.org/10.1162/99608f92.87b0ec26.

39. Chris O'Brien, "How NationBuilder's Platform Steered Macron's En Marche, Trump, and Brexit Campaigns to Victory," *VentureBeat*, July 14, 2017, https://venturebeat.com/2017/07/14/how-nationbuilder-helped-emmanuel-macron-secure-a-landslide-in-frances-legislative-elections/; Sue Halpern, "The Wild West of Online Political Operatives," *New Yorker*, October 1, 2019, https://www.newyorker.com/tech/annals-of-technology/the-wild-west-of-online-political-operatives.

40. Drew Olanoff, "Joe Green Steps Down as President of NationBuilder, Joins Andreessen Horowitz as EIR," *TechCrunch*, February 6, 2013, https://techcrunch.com/2013/02/06/joe-green-steps-down-as-president-of-nationbuilder-will-be-announcing-exciting-initiatives-in-the-coming-weeks/.

41. Jessica Guynn, "Mark Zuckerberg's Political Wingman: Fwd.us Founder Joe Green," *LA Times*, April 11, 2013, https://www.latimes.com/business/la-xpm-2013-apr-11-la-fi-tn-mark-zuckerberg-political-wingman-joe-green-20130411-story.html.

42. Halpern, "The Wild West of Online Political Operatives."

43. Samuel Axon, "96% of US Users Opt Out of App Tracking in iOS 14.5, Analytics Find," *Ars Technica*, May 7, 2021, https://arstechnica.com/gadgets/2021/05/96-of-us-users-opt-out-of-app-tracking-in-ios-14-5-analytics-find/.

44. Daniel Howley, "Apple Privacy Changes Hammer Social Media Stocks Beyond Meta," *Yahoo! Money*, February 3, 2022, https://money.yahoo.com/apple-privacy-changes-hammer-social-media-stocks-beyond-meta-162759059.html.

45. Howley.

46. Meta, "Introducing Meta: A Social Technology Company," Facebook press release, October 28, 2021, https://about.fb.com/news/2021/10/facebook-company-is-now-meta; Kyle Chayka, "Facebook Wants Us to Live in the Metaverse," *New Yorker*, August 5, 2021, https://www.newyorker.com/culture/infinite-scroll/facebook-wants-us-to-live-in-the-metaverse.

47. Tim Cook, "Is Apple's Privacy Push Facebook's Existential Threat," interviewed by Kara Swisher, *Sway*, April 5, 2021.

48. "Read Tim Cook's Interview with Chris Hayes and Kara Swisher," MSNBC, April 6, 2018, https://www.msnbc.com/msnbc/read-tim-cooks-interview-chris-hayes-and-kara-swisher-msna1087436 (full transcript from the "Revolution: Apple Changing the World" interview).

49. "Read Tim Cook's Interview with Chris Hayes and Kara Swisher." Apparently so incensed over the remarks, Zuckerberg ordered his management team to switch from iPhones to Android phones (though this was refuted in a Facebook blog post that explained, "We've long encouraged our employees and executives to use Android because it is the most popular operating system in the world." Facebook, "New York Times Update," Facebook press release, November 15, 2018, https://about.fb.com/news/2018/11/new-york-times-update/).

50. Ezra Klein, "Mark Zuckerberg on Facebook's Hardest Year, and What Comes Next," *Vox*, April 2, 2018, https://www.vox.com/2018/4/2/17185052/mark-zuckerberg-facebook-interview-fake-news-bots-cambridge.

51. Phil McCausland and Anna Schecter, "Cambridge Analytica Harvested Data from Millions of Unsuspecting Facebook Users," NBC News, March 17, 2018, https://www.nbcnews.com/news/us-news/cambridge-analytica-harvested-data-millions-unsuspecting-facebook-users-n857591.

52. McCausland and Schecter.

53. Meg Kelly and Elyse Samuels, "How Russia Weaponized Social Media, Got Caught and Escaped Consequences," *Washington Post*, November 18, 2019, https://www.washingtonpost.com/politics/2019/11/18/how-russia-weaponized-social-media-got-caught-escaped-consequences/.

54. Halpern, "The Wild West of Online Political Operatives."

55. "Transcript of Zuckerberg's Appearance before House Committee," *Washington Post*, April 10, 2018, https://www.washingtonpost.com/news/the-switch/wp/2018/04/11/transcript-of-zuckerbergs-appearance-before-house-committee/.

56. "Senate Hearing on Social Media Algorithms," CSPAN (video), April 27, 2021, https://www.c-span.org/video/?511248-1/senate-hearing-social-media-algorithms.

57. For a thorough and thoughtful account of the relationship between privacy and democracy, see Neil Richards, "Freedom," in *Why Privacy Matters* (New York: Oxford University Press, 2022), 131–163.

58. Salomé Viljoen, "Data Relations," *Logic(s)* 13, May 17, 2021, https://logicmag.io/distribution/data-relations/. See also Salomé Viljoen, "A Relational Theory of Data Governance," *Yale Law Journal* 131 (2021): 573–654.

59. Kate Kaye, "Cheat Sheet: Google Extends Cookie Execution Deadline Until Late 2023, Will Pause FLoC Testing in July," *DigiDay*, June 24, 2021, https://digiday.com/marketing/cheat-sheet-google-extends-cookie-execution-deadline-until-late-2023-will-pause-floc-testing-in-july/.

60. Justin Schuh, "Building a More Private Web," *Google: The Keyword* (blog), August 22, 2019, https://www.blog.google/products/chrome/building-a-more-private-web/.

61. Liam Quin, "Improving Web Advertising (by Changing the Web)," W3C, December 21, 2017, https://www.w3.org/community/web-adv/2017/12/21/improving-web-advertising-by-changing-the-web/#more-4.

62. Allison Schiff, "An Inside Look at the W3C with Strategy Lead Wendy Seltzer, as Debate Swirls around the Privacy Sandbox," *AdExchanger*, April 26, 2021, https://www.adexchanger.com/privacy/an-inside-look-at-the-w3c-with-strategy-lead-wendy-seltzer-as-debate-swirls-around-the-privacy-sandbox/.

63. "Notice of Intention to Accept Commitments Offered by Google in Relation to Its Privacy Sandbox Proposal," Competition & Market Authority Case 50972, June 11, 2021, https://assets.publishing.service.gov.uk/media/60c21e54d3bf7f4bcc0652cd/Notice_of_intention_to_accept_binding_commitments_offered_by_Google_publication.pdf.

64. Issie Lapowsky, "Concern Trolls and Power Grabs: Inside Big Tech's Angry, Geeky, Often Petty War for Your Privacy," *Protocol*, July 13, 2021, https://www.protocol.com/policy/w3c-privacy-war.

65. Lapowsky.

66. Lapowsky.

67. Competition and Markets Authority, "CMA to Investigate Google's 'Privacy Sandbox' Browser Changes," CMA press release, January 8, 2021, https://www.gov.uk/government/news/cma-to-investigate-google-s-privacy-sandbox-browser-changes.

68. Jonathan Mayer and Arvind Narayanan, "Deconstructing Google's Excuses on Tracking Protection," Freedom to Tinker, August 23, 2019, https://freedom-to-tinker.com/2019/08/23/deconstructing-googles-excuses-on-tracking-protection/.

69. Ben Savage (@btsavage), Twitter, July 9, 2021, 4:14 p.m., https://twitter.com/BenSava78155446/status/1413411126105329672; Ben Savage (@btsavage), Twitter, June 20, 2021, 4:45 p.m., https://twitter.com/BenSava78155446/status/1413411617530916865.

70. Savage (@btsavage), Twitter, June 20.

71. Bowdeya Tweh and Sahil Patel, "Google Chrome to Phase Out Third-Party Cookies in Effort to Boost Privacy," *Wall Street Journal*, January 14, 2020, https://www.wsj.com/articles/google-chrome-to-phase-out-third-party-cookies-in-effort-to-boost-privacy-11579026834.

72. "Heads of Facebook, Amazon, Apple, & Google Testify on Antitrust Law," CSPAN (video), July 29, 2020, https://www.c-span.org/video/?474236-1/heads-facebook-amazon-apple-google-testify-antitrust-law.

73. "Heads of Facebook, Amazon, Apple, & Google Testify on Antitrust Law."

74. "Facebook and Google Executives Testify on Competition and Privacy," CSPAN (video), September 21, 2021, https://www.c-span.org/video/?514791-1/facebook-google-executives-testify-competition-privacy.

75. "FTC Dark Patterns Workshop Transcript," FTC public events, April 29, 2021, https://www.ftc.gov/system/files/documents/public_events/1586943/ftc_darkpatterns_workshop_transcript.pdf.

76. "FTC Dark Patterns Workshop Transcript."

77. Daniel Solove and Woodrow Hartzog, "The FTC Zoom Case: Does the FTC Need a New Approach?," LinkedIn, November 11, 2020, https://www.linkedin.com/pulse/ftc-zoom-case-does-need-new-approach-daniel-solove.

78. Zoom Video Communications, Inc., Docket No. C-4731, File No. 192 3167 (Fed. Trade Comm'n 2020) (complaint), 8, https://www.ftc.gov/system/files/documents/cases/1923167zoomcomplaint_0.pdf.

79. Chris Hoofnagle (@hoofnagle), Twitter, November 9, 2020, 10:53, https://web.archive.org/web/20201109185824/https://twitter.com/hoofnagle/status/1325874277355147264.

80. Dissenting Statement of Commissioner Rebecca Kelly Slaughter, Zoom Video Communications, Inc., Docket No. C-4731, File No. 192 3167 (Fed. Trade Comm'n 2020), November 9, 2020, https://www.ftc.gov/system/files/documents/public_statements/1582918/1923167zoomslaughterstatement.pdf; Dissenting Statement of Commissioner Rohit Chopra, Zoom Video Communications, Inc., Docket No. C-4731, File No. 1923167 (Fed. Trade Comm'n 2020), November 6, 2020, https://www.ftc.gov/system/files/documents/public_statements/1582914/final_commissioner_chopra_dissenting_statement_on_zoom.pdf.

81. Sara Morrison, "Democrats Finally Get Their FTC Majority Back," *Vox*, May 11, 2022, https://www.vox.com/recode/23066131/alvaro-bedoya-ftc-confirmation-lina-khan.

82. Sally Wyatt, "Non-Users Also Matter: The Construction of Users and Non-Users of the Internet," in *How Users Matter: The Co-Construction of Users and Technology*, ed. Nelly Oudshoorn and Trevor Pinch (Cambridge, MA: MIT Press, 2003), 67–80.

83. Wyatt.

84. Nonuser, "TL;DR," accessed August 24, 2023, https://www.nonuse.org/home/#tldr.

85. Christine Satchell and Paul Dourish, "Beyond the User: Use and Non-Use in HCI," *Proceedings of the 21st Annual Conference of the Australian Computer-Human Interaction Special Interest Group: Design: Open 24/7* (2009): 9–16, https://doi.org/10.1145/1738826.1738829.

86. Blake E. Reid, "Internet Architecture and Disability," *Indiana Law Journal* 95 (2020): 591–647.

87. Liz Jackson, "Honoring the Friction of Disability," AIGADesign 2018, https://www.aiga.org/inspiration/talks/liz-jackson-honoring-the-friction-of-disability.

88. Sasha Costanza-Chock, *Design Justice: Community-Led Practices to Build the Worlds We Need* (Cambridge, MA: MIT Press, 2019). Sasha Costanza-Chock, a leading figure in the design justice movement, authored the book on design justice and titled a chapter "Nothing About Us Without Us," a phrase traced to international disability justice movements in the 1990s (see James I. Charlton, *Nothing About Us Without Us: Disability Oppression and Empowerment* [Berkeley: University of California Press, 1998]).

89. Michael Schudson, *Rise of the Right to Know: Politics and the Culture of Transparency, 1945–1975* (Cambridge, MA: Harvard University Press, 2015), 37.

90. Schudson, 29.

91. Schudson, 63.

92. Margaret B. Kwoka, *Saving the Freedom of Information Act* (New York: Cambridge University Press, 2021).

93. Globally influential in a way the Swedish 1766 law and Finnish 1951 law were not; Schudson cites National Security Archive director Thomas Blanton for this term and crediting FOIA as its origin source. The term is also used in a 2014 article describing concern from executives that transparency has gone "too far" in demands for accountability. "The Openness Revolution," *The Economist*, December 11, 2014, https://www.economist.com/business/2014/12/11/the-openness-revolution.

INDEX

Abacus Direct, 139
Ackermann, Kevin, 86
Acquisti, Alessandro, 155
AdLINK, 122
AdSense, 146
Advertising Age, 121
Advertising: A New Weapon in the World-Wide Fight for Freedom; A Guide for American Business Firms Advertising in Foreign Countries, 125
Advertising as infrastructure, 117–124
Allen, Anita, 154–155
Allen, Paul, 202n11
Alternative networks, 19
Altman, Irwin, 13, 15
Amazon, 135, 148, 184–185
American Challenge, The (Servan-Schreiber), 27, 35
American Digital Equipment Corporation (DEC), 77
American Express, 132
American Federation of Information Processing Societies, 39
America Online (AOL), 1, 8, 73, 92–93, 111, 145, 224n91
America Online Simplified (Maran), 224n91
America's Uncounted People, 40

Andreessen, Marc, 109–111, 136
Anonymity-after-authorization, 76
Apple, 11, 69, 70, 92, 184
  iPhone privacy updates and, 177–182
  photos stored by, 172
  Safari browser, 148
Apple HyperCard, 108
App Tracking Transparency (ATT) program, 177
ARPANET, 8, 73, 75–78
  Requests for Comments (RFC) filed regarding, 76
Assault on Privacy: Computers, Data Banks, and Dossiers, The (Miller), 39
Association for Computing Machinery (ACM), 48
Association of Data Processing Service Organizations (ADAPSO), 97
AT&T, 93

Badham, John, 85
Bailey, William, 53
Baker, Kenneth, 208n95
Bannon, Steve, 181
Baran, Paul, 38–39, 75
Barksdale, Jim, 136
Basic Law for the Federal Republic of Germany, 14, 57, 88–89

Bedoya, Alvaro, 186
Behlendorf, Brian, 113–117, 120, 137, 231–232n46
Belgium, 61, 171–172
Berger v. United States, 96
Berman, Jerry, 144
Berners-Lee, Tim, 105, 107, 108–109, 110, 112–113, 135
Better Business Bureau (BBB), 140
Beyda, Gil, 118
Big data, 179
Bildschirmtext (BTX), 87–91
Bina, Eric, 109–111
Bing, Jon, 65
Biometric Information Privacy Act (BIPA), Illinois, 11, 172
Birkett, Norman, 78
Birkett Committee, 78–79
Blackburn, Marsha, 185
Bonome v. Kaysen, 14
Boston Globe, 93
Bouk, Dan, 194n26
Boundary management, 13, 15
Brandeis, Louis, 13, 36, 45, 95–96
Branscomb, Chuck, 31
Brexit, 179
British Computer Society (BCS), 46–48
British National Archives, 20
British Tabulating Machine (BTM) Company, 28
Bull, 28, 31–32, 61, 70
Bulletin Board Systems, 8, 73, 85–91
Burroughs, 29
Bush, George H. W., 133
Bush, George W., 161
Bushell, Colleen, 110–111

Cailliau, Robert, 106–107, 108, 112
California Consumer Privacy Act (CCPA), 7, 175–176
Cambridge Analytica, 5, 180–181
Campbell-Kelly, Martin, 20

Canada
 computer industry in, 34
 computer laws in, 62, 65
Case, Stephen, 92
Catlett, Jason, 144–145, 148
Center for Democracy and Technology (CDT), 144
Centre d'Etudes Pratiques d'Informatique et d'Automatique (CEPIA), 51
Chaos Computer Club (CCC), 85–91
Charter of Fundamental Rights of the European Union, 15–16, 149–150, 166
Choosers, 16–17
Chrome browser, 160, 169
CII, 35
Clark, Jim, 110–112
Clearview AI, 172–173
Click-throughs, 2
Clinton, Bill, 9, 106, 122, 123–124, 133
CNET, 77
Code of Fair Information Practices, 42–43, 46
Code of Fair Labor Practices, 42
Cohen, Julie, 17, 154–155
Cohen, Lizabeth, 125
Coleman, Gabriella, 74
Commissariat à l'énergie atomique, 32
Commission nationale de l'informatique et des libertés (CNIL), 60–61, 63, 171
Commission on Data Processing and Freedom, 51–52
Committee for Liquidation or Subversion of Computers (CLODO), 85
Committee of Experts on Data Protection, 64
Commodore computers, 69, 92, 99
Compagnie internationale pour l'informatique (CII), 32

Compaq, 70
CompuServe, 93, 100, 111
Computer Analysts and Programmers (CAP), 46
Computer characters, 6–8, 10–11, 189–190
  tech governance with new, 186–189
Computer Fraud and Abuse Act (CFAA), 98–100, 141
  DoubleClick and, 142–143
Computer industry, 26–36, 201–202n20
  advertising as infrastructure in, 117–124
  American, 27–28, 35–36, 52–53
  British, 32–33, 46–49
  Canadian, 34
  early, 27–29
  French, 31–32, 35, 51–52
  German, 33–36, 55–57
  hypertext and, 107–109
  Japanese, 34, 36
  post-war market growth and, 34–35
  power dynamics of, 25–26
  punched-card machine, 27–30
  reports on privacy and, 36–54
  Swedish, 34, 50–51
  transborder data flows and, 63–71
Computerization of Society project, 52
Computer Power and Human Reason (Weizenbaum), 40
Computers and Privacy White Paper, 49
Computer Weekly, 152
Computerworld, 132
Consent, 2–5, 23, 193n18, 247n60
  as consent theater, 3, 170–173
  data privacy and, 141–142
  data subjects and, 6–8
  Do Not Track (DNT) consent mechanism, 155–160
  as healthy competition, 182–186
  new data subject and, 173–176
  pop-ups, 11
  as privacy civics, 177–182
  tools for, 154–160
  users vs. consumers and, 5–6
Consent dilemma, 4
Consent theater, 3, 170–173
"Consumer Bill of Rights," 126
Consumer Privacy Protection Act (CPPA), 145
Consumers, 5–6
  as choosers, 16–17
  data protection for, 22
  privacy (*see* Privacy consumers)
Consumer societies and consumerism, 124–129
Control as privacy, 14–15, 37
Control Data Corporation (CDC), 29, 32
Controlling the Assault of Non-Solicited Pornography and Marketing (CAN-SPAM) Act, 145–146
Control of Personal Information Bill 1971, 208n95
Convention for the Protection of Individuals with Regard to Automatic Processing of Personal Data (Convention 108), 64–66
Convention on Human Rights, 149
Cook, Tim, 180, 182
Cookie Central, 134
Cookies
  alternative networks and, 19
  court cases and legislative action related to, 142–148
  data privacy issues and, 141–142
  defined, 1
  EU directive on, 148–154
  hypertext transfer protocol (HTTP) and, 113–117
  increase in use of, 2
  as memory system, 2
  opt-out and, 134–135
  periodization and, 20
  privacy tools for, 154–160
  retirement of, 11, 169, 170, 183

Cookies (cont.)
  scholarly attention on history and
    interventions on, 21–24
Cooley, Thomas M., 95
Cool Site of the Day (CSotD), 120–121
Coons, Christopher, 181
Cophra, Rohit, 186
Cortada, James, 30
Council of Europe, 63–65, 149
Court of Cassation, 84
Cox, Donna, 110
Crain, Matt, 146
Cranor, Lorrie, 155
Credit bureaus, 70, 128–129
Cuckoo's Egg, The (Stoll), 90
Customer proprietary network
  information (CPNI), 165
Cyclades, 77–78

D'Agapeyeff, Alex, 46, 48
Dark patterns, 185
Data banks, 39
Databanks in a Free Society project, 36
*Database of Dreams: The Lost Quest to
  Catalog Humanity* (Lemov), 205n49
Datagrams, 77–78
Data Inspection Board (DIB), Swedish,
  51, 57–58
Data networks, 74–80
Data privacy, 141–142
  boundary management and, 13, 15
  Code of Fair Information Practices
    and, 42–43, 46
  codified in Europe in the 1800s and
    early 1900s, 219–220n49
  consumerism and, 124–129
  control as, 14–15, 37
  for cookies, building of, 154–160
  customer proprietary network
    information (CPNI), 165
  dark patterns in, 185
  Do Not Track (DNT) and, 155–160,
    175–176

  early calls for digital, 208n94
  fair information practices (FIPs), 22
  Freedom of Information Act (FOIA)
    and, 42–43, 59, 188, 189–190
  legal construction of technology and,
    17–21
  opt-out and, 129–140, 174–175,
    178–179
  Privacy Shield and, 164–165
  public concerns over, 44–45
  Safe Harbor principles and, 10, 160–
    161, 163
  Terrorist Surveillance Program (TSP)
    and, 161–163
  transatlantic breakdown in, 160–167
Data Protection (DP) Directive, EU, 67–
  68, 71, 154, 167
Data Protection Act, Germany, 55–56
Data Protection Act, United Kingdom,
  62–63
Data Protection Committee, UK, 49
Data protection impact assessments
  (DPIA), 174
Data protectionism, 63
Data subject(s), 6–8, 190, 194–195n27
  HEW Report on, 44
  new conceptualization of, 173–176
  not embraced in the United States,
    70
  origins of, 25, 37, 39
  transborder data flows and, 68
Data Surveillance Bill 1969, 208n95
Davies, Donald, 75, 77, 78
Davis, Glenn, 120
De Gaulle, Charles, 32
Denmark, 62
Department of Commerce (DOC), 106
D'Estaing, Giscard, 52, 60
Deutsch Hollerith Maschinen GmbH
  (Dehomag), 27, 28
Digital counter public, 88
Direct Mail Advertising Association
  (DMAA), 130–131, 139–140

INDEX

Direct Marketing Association, 121, 139–140
Dissemination of information, 13
Dobbs, Guy, 40–41
Doctorow, Cory, 3
Do Not Call Registry, 155
Do Not Track (DNT), 155–160
  California Consumer Privacy Act (CCPA) on, 175–176
Dot-com crash, 9
Doty, Nick, 158
DoubleClick, 118, 119, 121–122, 134–136, 139–140
  investigations of, 140, 142–144
  opt-out option for, 156
  purchased by Google, 146
Driscoll, Kevin, 81, 86
Dumb terminals, 92
Dunn, Edgar S., Jr., 37, 38
Dunn Report, 37

Earthlink, 1
Electronic Communications Privacy Act (ECPA), 96–98, 141
  DoubleClick and, 142–143
Electronic Frontier Foundation (EFF), 161–162
Electronic Mail Association, 97
Electronic Numerical Integrator and Computer (ENIAC), 29
Elias, Howard, 69
ELIZA program, 40
Engage, 119, 122
Engelbart, Doug, 107
English Electric Leo Marconi (EELM), 32
Erdos, David, 149
Ervin, Samuel, Jr., 39, 53
EUNet, 8
Euronet, 73, 100–103, 228n140
European Academic and Research Network, 228n140
European Convention on Human Rights (ECHR), 63, 79

European Court of Human Rights, 65, 89
European Court of Justice (ECJ), 171
European Economic Community, 35, 67, 126
European Informatics Network (EIN), 78, 228n140
European Organization for Nuclear Research (CERN), 105, 106–107, 109
European Union
  Charter of Fundamental Rights of the, 15–16, 149–150, 166
  cookie directive, 148–154
  Data Protection (DP) Directive, 67–68, 71, 154, 167
  ePrivacy Directive, 7, 9–10, 102, 149–150, 153
  General Data Protection Regulation (GDPR), 11, 149–150, 166–167, 170
  Treaty on European Union (Maastricht Treaty), 67
  Treaty on the Functioning of the European Union (TFEU), 126–127
European Unix Network, 228n140
EU–US Safe Harbor arrangement, 10, 160–161, 163
Experimental Packet Switched Service (EPSS), 78

Facebook, 5, 147, 148, 163, 184–185
  bashing of Apple by, 10, 177
  Cambridge Analytica scandal and, 5, 180–181
  compliance with laws in France, 171
  rebranded as Meta, 179–180
  sharing of consumer data by, 12, 23–24
Fair Credit Reporting Act (FCRA), 70, 129
Fair information practices (FIPs), 22, 199n85

Family Educational Rights and Privacy Act, 70
Federal Communications Commission (FCC), 91, 133–134
Federal Register, 59
Federal Trade Commission (FTC), 9, 91, 106, 127, 129, 132–133, 138, 140
  on circumventing browser privacy settings, 185–186
  on dark patterns, 185
  Do Not Call Registry and, 155
  Do Not Track (DNT) consent mechanism, 155–156
  online privacy reports issued by, 144–145
  power and jurisdiction of, 145–147, 165–166
Ferranti, 32, 34, 78
FidoNet, 87
51Degrees, 183
File Retrieval and Editing SyStem (FRESS), 108
Financial Times, 134
Firefox, 157, 169
Flaherty, David, 65
Ford, Gerald, 58
Foreign Intelligence Surveillance Amendments Act (FISA), 161–163
1401 computer, 30–32, 33
France
  anonymous users on Minitel in, 80–85
  Bull, 28, 31–32, 61, 70
  Centre d'Etudes Pratiques d'Informatique et d'Automatique (CEPIA), 51
  Commissariat à l'énergie atomique, 32
  Commission nationale de l'informatique et des libertés (CNIL), 60–61, 63, 171
  Committee for Liquidation or Subversion of Computers, 85
  Compagnie internationale pour l'informatique (CII), 32
  computer industry in, 31–32, 35, 51–52
  computer laws in, 60–61
  Court of Cassation, 84
  Institut de Recherche en Informatique et en Automatique (IRIA), 51
  Minitel, 8, 61, 72, 219n41
  Plan Calcul, 51, 60
  Postal and Electronic Communications Code, 82
  Prestel, 8, 73, 81, 219n41
  On Reinforcing the Guarantees of Individual Rights of Citizens, 1970, 84
  Réseau à Commutation de Paquets (RCP), 77
  Système Automatisé pour les Fichiers Administratifs et Répertoires des Individus (SAFARI), 51
  transborder data flows in, 67–68
  Tricot Commission, 52, 60
France Telecom, 81–82
Freedom of Information Act (FOIA), 42–43, 59, 188, 189–190
Fujitsu, 36
Full Frontal with Samantha Bee (TV show), 172

Gajda, Amy, 14
Gallagher, Cornelius E., 38, 128, 129, 130
Gaps in Technology report, 35
Garcia-Swartz, Daniel, 20
Gates, Bill, 136, 202n11
Genachowski, Julius, 157
General Data Protection Regulation (GDPR), 11, 149–150, 166–167, 170
  data protection impact assessments (DPIA) and, 174
General Electric, 29, 32

INDEX

Genome Database, 116
Gen-O-Pak, 128
Georges, Marie, 51, 60, 68
Germany
    Basic Law for the Federal Republic of Germany, 14, 57, 88–89
    Bulletin Board Systems, 8, 73, 85–91
    CCC hackers against BTX in, 85–91
    computer industry in, 33–36
    computer laws in, 55–57
    Deutsch Hollerith Maschinen GmbH (Dehomag), 27, 28
    Klass v. Germany, 89
    Nixdorf Computer AG, 34, 36, 70
    Siemens, 33–36, 70
    Telemedia Act, 170–171
    transborder data flows in, 68
Goldwater, Barry, Jr., 53
Google, 171, 184–185
    accused of privacy gaslighting, 184
    Chrome browser, 160, 169
    circumventing user attempts to block cookies, 148
    opt-out option offered by, 156
    photos stored by, 172
    purchase of DoubleClick by, 146
    retirement of cookies announced by, 11, 169, 170, 183
Google Advertising ID (GAID), 177
Gopher services, 109, 110
Gore, Al, 123
GRAMMATRON, 120, 155
Green, Joe, 179
Grobe, Michael, 109
Guidelines on the Protection of Privacy and Transborder Flows of Personal Data, 64

Hackers, 74, 80, 85
    CCC, against BTX in Germany, 85–91
Hacker's Dictionary, The, 74
Hacking, Ian, 6
Hall, Wendy, 108

Hanover Hackers, 89–90
Hardie, Ted, 135–136
Harry, Prince, 172
Hartzog, Woodrow, 186, 193n18
Hedlund, Marc, 115
Heidinger, Willy, 27
Hessen Datenschutzgesetz, 55
Hewlett Packard, 70
HEW Report, 36, 39–44, 49, 54, 58, 129, 138
    on the Data Protection Act in Germany, 55–56
    on mailing lists, 131
Hicks, Mar, 32–33
Hi-media, 122
Hollerith, Herman, 27
Holtman, Koen, 115–116
Hondius, Frits, 64, 65
Honeywell, 29, 32, 35
Hoofnagle, Christopher, 127, 133, 186
HotWired.com, 113, 231n46
How Our Days Became Numbered (Bouk), 194n26
HTTP Working Group, 135–136
Huckfield, Leslie, 208n95
Hypertext, 107–109
Hypertext transfer protocol (HTTP), 112–117

IBM, 33, 69
    early years of, 27–29, 34, 99
    foreign operations subsidiary of, 30
    1401 and System/360 computers, 30–32, 33, 34
    market dominance of, 34–35, 36
    personal computer market and, 69–70
    privacy issues addressed by, 41–42
Prodigy, 93
technical working group, 49
technological solutionism and, 43
in West Germany after World War II, 33

If Then: How the Simulmatics Corporation Invented the Future (Lepore), 256n38
Immortal Life of Henrietta Lacks, The, 14
Information collection, 13
Information frames, 80–81
Information processing, 13
In re Pharmatrak, Inc. Privacy Litigation, 147–148
Institut de Recherche en Informatique et en Automatique (IRIA), 51
Institute for Advanced Study, 38
Interactive Advertising Bureau, 150
Interactive Marketing Group (IMG), 121
Interception of Communications Act, 80
International Computers and Tabulators (ICT), 32
International Computers Limited (ICL), 32, 70
Internet, development of the, 68–69, 228n140–141
 browsers in, 109–112
 early commercial websites in, 231–232n46
 hypertext in, 107–109
 hypertext transfer protocol (HTTP) in, 112–117
Internet Advertising Bureau (IAB), 150
Internet Advertising Federation, 117
Internet Engineering Task Force (IETF), 20, 136
Internet Research Agency, 181
Iowa Law Review, 68

Japan
 computer industry in, 34, 36
 Nippon Electric Company (NEC), 32
Jargon File, 74
Jasanoff, Sheila, 17–18
Jaye, Daniel, 119
Jennings, Tom, 87
Jobs, Steve, 107
Joinet, Louis, 51, 60, 65

Kallet, Arthur, 127
Kaminsky, Dan, 156
Kaminsky, Margot, 176
Kaphan, Shel, 135
Katz v. United States, 96
Kaysen, Carl, 38
Kaysen Report, 38
Kennedy, John F., 126, 179
Kerr, Donald, 162–163
Keypunch machines, 29
Khanna, Ro, 23
Kirby, Michael, 65
Klass v. Germany, 89
Koch, Ed, 53, 54
Kristol, Dave, 135, 136
Kwoka, Margaret, 188, 189–190

Lacks, Henrietta, 14
Lanphere, Patricia, 40
Lauer, Josh, 129
Laws, computer, 55–63
 Canadian, 62
 consumer privacy, 124–129
 French, 60–61
 German, 55–57
 opt-out, 129–140
 related to cookies, 145–148
 Swedish, 57–58
 transborder data flow, 63–71
 United Kingdom, 62–63, 150
 United States, 58–60, 96–100, 127–128
 wiretapping, 79–80, 96–98
Le Défi Américain (Servan-Schreiber), 27
Legal construction of technology, 17–21
Leibowitz, Jon, 156
Lemoine, Philippe, 51–52
Lemov, Rebecca, 205n49
Lepore, Jill, 256n38
Lerach, Bill, 142, 242n2

Le Secret Des Fichiers (Gallouédec-
 Genuys and Maisl), 52
Libération, 85
Life insurance industry, 194n26
Lincoln, Abraham, 95
Lindop, Norman, 49
Lindop Report, 49, 62
Link, Terry, 172
Linowes, David, 53, 59
London Times, 44
Long, Billy, 24
Los Angeles Times, 94
Lotus MarketPlace, 119
Lowenthal, Alan, 159–160
Lurking: How a Person Became a User
 (McNeil), 73
Lusers, 74, 91–100
Luxembourg, 62
Lynx, 109

Maastricht Treaty (Treaty on European
 Union), 67
Macron, Emmanuel, 179
Mactaggart, Alastair, 174–175
Mailland, Julien, 81
Mail Preference Service, 130–131
"Making Up People" (Hacking), 6
Malone, James, 89
Malone v. UK, 79
Maran, Ruth, 224n91
Markle, Meghan, 172
Marrinan, Patrick, 78
Marshall Plan, 75, 125
Matra, 61
Mayer, Jonathan, 157, 184
McCain, John, 144, 148
McDonald, Aleecia, 155, 157–160
McNeil, Joanne, 73
Memex machine, 107–108, 112
Merriman, Dwight, 117–118
Meta, 179–180. *See also* Facebook
Metromedia, 132

Mhyrvold, Nathan, 136
Microcosm, 108
Microsoft, 136–137, 146, 156
 Internet Explorer browser, 148
 opt-in option offered by, 158
Millard, Wenda Harris, 118
Miller, Arthur R., 37, 39, 41, 128
Miller's Court (TV show), 39
Minc, Alain, 52
Minitel, 8, 61, 72, 91, 98–99, 219n41
 anonymous users on, 80–85
Minitel: Welcome to the Internet (Mailland
 and Driscoll), 81
Minnesota Fair Information Practices
 Act, 53
MinTech, 32
MIT, 39
Mitterrand, François, 61
Montulli, Lou, 21, 109–110, 112–114,
 135, 137
Moorhead, Carlos, 97
Moral magic, 4
Morgan, Dave, 118–119
Mosaic archive, University of Illinois,
 20
Mosaic browser, 109–112
Moss, John, 188
Mozilla, 157
Muslim Pro, 182

NAACP, 179
Naked Society, The (Packard), 38
Napster, 179
Narayanan, Arvind, 157, 184
Nash, Edward, 118
National Academy of Sciences, 36
National Cash Register Company
 (NCR), 29
National Center for Supercomputing
 Applications (NCSA), 109–110
National Data Center, 37, 38, 75
National Do Not Call Registry, 134

National Information Infrastructure (NII) plan, 123
National Physical Laboratory (NPL), 75, 78
National Science Foundation Network (NSFNET), 109
National Security Agency (NSA), 161–162
NationBuilder, 179
Negative check off system, 9
Nelson, Jonathan, 231n46
Nelson, Matthew, 231n46
Nelson, Ted, 108
Netflix, 184
Netscape Communications, 111
Netscape Navigator, 2, 21, 111–112, 117
    cookie settings, 134, 136–138
Network Advertising Initiative (NAI), 139–140, 156
New Yorker, 130
New York Times, 69, 108, 112, 120, 121, 161, 175, 177
Nippon Electric Company (NEC), 32
Nisenholtz, Martin, 121
Nissenbaum, Helen, 154–155
Nixdorf, Heinz, 34
Nixdorf Computer AG, 34, 36, 70
Nixon, Richard, 39, 52–53, 58, 179
Noble, Safiya, 172
Nora, Simon, 52
Nora-Minc Report, 52
Nudge (Thaler and Sunstein), 16

Obama, Barack, 23, 162, 164, 179
O'Connor, Kevin, 117–118, 121–122
Olivetti, 31
Olmstead v. United States, 96
100,000,000 Guinea Pigs (Kallet and Schlink), 127
Online Privacy Alliance, 145
On Reinforcing the Guarantees of Individual Rights of Citizens, 1970, 84
Opt-out, 129–140, 174–175, 178–179
Organization for Economic Cooperation and Development (OECD), 35, 62, 64–65

Packard, Vance, 38
Packet Switch Stream (PSS), 78
Pai, Ajit, 166
Parker, Sean, 179
Parliamentary Commission on Publicity and Secrecy Legislation (OSK), 50–51
Parsons, Carole, 40–41, 52–53
Patrick, Neal, 99
Pellow, Nicola, 107
Pelosi, Nancy, 162
"People Machine," 178–179
People magazine, 112
Periodization, 20
Personal computers, 69–71
    hobbyists and, 73–74
Personality rights, 14
Pichai, Sundar, 185
Pitofsky, Robert, 133, 138–139, 145
Plan Calcul, 51, 60
Planet49 case, 171
Politics
    Cambridge Analytica scandal and, 5, 180–181
    use of opt-out in, 178–179
Polonetsky, Jules, 144
Pop-ups, 2
Portugal, 62, 63
    transborder data flows in, 67
Postal, telegraph, and telephone corporations (PTTs), 76–77, 101
Postal and Electronic Communications Code, 82
Pouzin, Louis, 78
Power dynamics of computing, 25–26
Powers, James, 27
Powers-Samas, 28, 32

Powers Tabulating Machine Company, 27
Prestel, 8, 73, 81, 87, 91, 219n41
Privacy Act of 1974, 53, 58–60
Privacy and Computers report, 62
Privacy and Freedom (Westin), 36
Privacy civics, consent as, 177–182
Privacy consumers, 6, 8–9, 105–106
  advertising as infrastructure and, 117–124
  consent as healthy competition for, 182–186
  hypertext and, 107–109
  opt-out and, 129–140
  self-regulation by, 138, 140
  stateless web and, 106–117
Privacy-enhancing technologies (PETs), 9, 155
Privacy Interest Group (PING), 183–184
Privacy Online report, 138
Privacy policies, 3–4, 234n88
  European Convention on Human Rights (ECHR), 63
  reports on, 36–54
  United States' reputation for lackluster, 36
Privacy Protection Study Commission (PPSC), 53–54, 59–60
  on mailing lists, 131–132
Privacy Sandbox, 183
Privacy Shield, 164–165
Private Lives and Public Surveillance (Rule), 66
Prodigy, 93–94, 111
Projectories, 10–11
Prosser, William, 13
Protectionism, 23
Punched-card machines, 27, 28–30

Quantum Computer Services, 92

Rand Corporation, 38, 39, 75
RCA, 29, 32
Reagan, Ronald, 133
Real Media, 118, 119, 122
Real Time Club, 78
Records, Computers and the Rights of Citizens, Report of the Secretary's Advisory Committee on Automated Personal Data Systems. See *HEW Report*
Regan, Priscilla, 66
Reidenberg, Joel, 66, 68
Remington Rand, 27, 29, 32, 33, 34
Reports, computer, 36–54
  American vs. European privacy policies examined in, 36–37
  Association for Computing Machinery (ACM), 48
  British Computer Society (BCS) and, 46–48
  Code of Fair Information Practices and, 42–43, 46
  Computers and Privacy White Paper, 49
  Data and Privacy Report, OSK, 50–51
  on data banks, 39
  data subject in, 37, 39
  HEW Report, 36, 39–44, 49, 54, 58, 129, 131, 138
  Lindop Report, 49, 62
  National Data Center and, 37, 38
  Nora-Minc Report, 52
  Privacy Protection Study Commission (PPSC), 53–54, 131–132
  Younger Report, 20, 42, 44–46, 48–49, 62
Réseau à Commutation de Paquets (RCP), 77
Revenues, ad, 117–124
Rezac, Charles, 109
Richards, Neil, 193n18
Richardson, Elliot, 39, 44, 52
Right of printout, 48
"Right to Privacy, The" (Warren and Brandeis), 36
Robbins, Robert, 116

"Robots Will Set Us Free to Serve," 44
Rockefeller, Jay, IV, 156
Rosewell, James, 183
Ruggles, Richard, 37
Ruggles Report, 37
Rule, James B., 66
Russell, Andrew, 77

Saab D21 mainframe, 34
Safari browser, 148, 169
Safe Harbor principles, 10, 160–161, 163
Savage, Ben, 184
Schlink, F. J., 127
Schudson, Michael, 14, 188
Schultheiß, Michael, 122
Schwartz, Paul, 56, 66, 68, 94
Sears, 93
Secure Sockets Layer (SSL) protocol, 111
Self-management, privacy, 4
Self-regulation, 138, 140
Servan-Schreiber, Jean-Jacques, 27, 35
Shrems, Max, 163–165
Sieghart, Paul, 49
Siemens, 33–36, 70
Siino, Rosanne, 112
Silicon Graphics (SGI), 110–112
Simitis, Spiros, 55–56, 65, 67, 68
Sink, Eric, 135
60 Minutes, 184
Skolnick, Cliff, 231n46
Slaughter, Rebecca, 186
Snapchat, 172
Snowden, Edward, 160–161
Social Dilemma, The (film), 184
Society of Certified Data Processors (SCDP), 41
Sociotechnical imaginaries, 10
Soghoian, Chris, 156
Solove, Daniel, 13, 185–186
Soltani, Ashkan, 175
Sony, 152–153

Source, The, 93
Spain, 62, 63
  transborder data flows in, 67
Spitzer, Eliot, 140
Stamm, Sid, 157
Stateless web, 106–117
Stearns, Peter, 124
Steiger, Janet Dempsey, 133
Stoll, Cliff, 89
Stored Communications Act (SCA), 98
Strickland, Karl, 80
Stustch, Michael, 122
Sunstein, Cass, 16
Surveillance capitalism, 169–170, 173
Sweden
  computer industry in, 34, 50–51
  computer laws in, 57–58
  Data Inspection Board (DIB), 51, 57–58
  Parliamentary Commission on Publicity and Secrecy Legislation (OSK) in, 50–51, 57
  Swedish Board for Computing Machinery, 34
  Swedish Data Act, 204n45
Swedish Board for Computing Machinery, 34
Swedish Data Act, 204n45
Swicher, Kara, 23, 180
Swindle, Orson, 145
Switzerland, 62
System/360 computer, 30–31, 33, 34
Système Automatisé pour les Fichiers Administratifs et Répertoires des Individus (SAFARI), 51

Taft, William Howard, 96
Tandy, 69
Tapping. *See* Wiretapping
Technological solutionism, 43
Telecommunications Act of 1934, 165

Telecommunications Act of 1996, 222n78
Telemedia Act, Germany, 170–171
Telephone Consumer Protection Act (TCPA), 133–134
Telic-Alcatel, 61
Telidon, 91, 219n42
Terrorist Surveillance Program (TSP), 161–163
Thaler, Richard, 16
TikTok, 172
Time magazine, 112
Tools, privacy, 154–160
Tracking Protection Working Group (TPWG), 155, 157–158
Transborder data flows, 63–71
  Convention for the Protection of Individuals with Regard to Automatic Processing of Personal Data (Convention 108), 64–66
  Council of Europe on, 63–64
  Data Protection (DP) Directive, 67–68, 70
  Organization for Economic Cooperation and Development (OECD) on, 64–65
Transpac, 77
Transparency and Consent Framework (TCF), Belgium, 171–172
Treaty of Lisbon, 166
Treaty of Rome, 66
Treaty of Washington, 95
Treaty on European Union (Maastricht Treaty), 67
Treaty on the Functioning of the European Union (TFEU), 126–127
Tricot, Bernard, 51–52
Tricot Commission, 52, 60
Trump, Donald, 165, 179, 181
TRUSTe, 156
Turow, Joseph, 155

Twitter, 173, 179
Tyranny of choice, 16

Unilateral globalism, 122
United Kingdom, the
  authorized users and data networks in, 74–80
  Birkett Committee, 78–79
  British Computer Society (BCS), 46–48
  British National Archives, 20
  British Tabulating Machine (BTM) Company, 28
  computer industry in, 32–33, 46–49
  computer laws in, 62
  Data Protection Act 1981, 62–63
  Data Protection Committee, 49
  Ferranti computers, 32, 34, 78
  Interception of Communications Act, 80
  International Computers Limited (ICL), 32, 70
  Prestel, 8, 73, 219n41
United States, the
  advertising as infrastructure in, 117–124
  ARPANET, 8, 73, 75–76
  Berger v. United States, 96
  Biometric Information Privacy Act (BIPA) in, 11, 172
  California Consumer Privacy Act (CCPA), 7, 175–176
  Computer Fraud and Abuse Act (CFAA), 98–100, 141–143
  computer industry in, 27–28, 35–36, 52–53
  computer laws in, 58–60
  "Consumer Bill of Rights" in, 126
  consumerism in, 124–129
  credit bureaus in, 70, 128–129
  data subject in, 70

United States, the (cont.)
  Direct Mail Advertising Association (DMAA) in, 130–131
  Electronic Communications Privacy Act (ECPA), 96–98
  Federal Trade Commission (FTC), 9, 91, 106, 127
  Fourth Amendment, 94–95
  Freedom of Information Act (FOIA), 42–43, 59, 188, 189–190
  Katz v. United States, 96
  luser screen names in, 91–100
  Olmstead v. United States, 96
  opt-out in, 129–140
  Postal Service Act, 1792, 94–95
  Privacy Act of 1974, 58–60
  Privacy Protection Study Commission (PPSC), 53–54, 59–60
  Stored Communications Act, 98
  wiretapping laws in, 96–97
Unix machines, 110, 114
Usenet, 73, 92, 100, 120
Users, 5–6
  anonymous, on Minitel, 80–85
  authorized, 74–80
  CCC hackers against BTX in Germany, 85–91
  consent as privacy civics and new, 177–182
  distinguished from hackers, 74
  Euronet, 100–103
  legal definition of, 74
  luser screen names, 91–100
  origin story of, 72
US-EU Safe Harbor Framework, 147

Van Dam, Andries, 108, 120
Veale, Michael, 173
Video Privacy Protection Act, 70
Videotex, 80–85, 91, 93–94
Viljoen, Salomé, 182
Vincent, David, 13
Virtual circuits, 77–78

Walden, Greg, 23–24
Waldman, Ari, 173
Walker, Philip M., 97
Wall Street Journal, 2, 93, 119, 177
Ware, Willis H., 39, 41, 45–46, 53
WarGames (film), 85, 99
Warren, Samuel D., II, 13, 36, 45, 95
Washington Post, 175, 177
Watson, Thomas J., Sr., 29, 43
Webonomics, 120
Weiss, Melvyn, 142
Weizenbaum, Joseph, 39–40, 41, 44
Western Union, 95, 128–129
Westin, Alan, 14, 36, 37, 128
Whalen, David, 134
Wheeler, Tom, 165–166
Wheeler-Lea Act, 127
Whitman, James, 18–19
Wilson, Harold, 32
Wired magazine, 1, 113, 120
Wiretap Act, 1968, 97–98
  DoubleClick and, 142–143
Wiretapping, 79–80, 96–98
Woods, Neil, 80
World Wide Web Consortium (W3C), 20, 109, 113, 135, 155–160, 175
  Improving Web Advertising Business Group, 183
  Privacy Interest Group (PING), 183–184

Xerox, 40
X-Mosaic, 110

Younger, Kenneth, 45
Younger Report, 20, 42, 44–46, 48–49, 62

Zimmermann, Cyril, 122
Zoom, 185–186
Zuckerberg, Mark, 3, 5–6, 12, 23–24, 179–181
Zuse, Konrad, 33

Information Policy Series

Edited by Sandra Braman

Virtual Economies: Design and Analysis, Vili Lehdonvirta and Edward Castronova

Traversing Digital Babel: Information, E-Government, and Exchange, Alon Peled

Chasing the Tape: Information Law and Policy in Capital Markets, Onnig H. Dombalagian

Regulating the Cloud: Policy for Computing Infrastructure, edited by Christopher S. Yoo and Jean-François Blanchette

Privacy on the Ground: Driving Corporate Behavior in the United States and Europe, Kenneth A. Bamberger and Deirdre K. Mulligan

How Not to Network a Nation: The Uneasy History of the Soviet Internet, Benjamin Peters

Hate Spin: The Manufacture of Religious Offense and Its Threat to Democracy, Cherian George

Big Data Is Not a Monolith, edited by Cassidy R. Sugimoto, Hamid R. Ekbia, and Michael Mattioli

Decoding the Social World: Data Science and the Unintended Consequences of Communication, Sandra González-Bailón

Open Space: The Global Effort for Open Access to Environmental Satellite Data, Mariel Borowitz

You'll See This Message When It Is Too Late: The Legal and Economic Aftermath of Cybersecurity Breaches, Josephine Wolff

The Blockchain and the New Architecture of Trust, Kevin Werbach

Digital Lifeline? ICTs for Refugees and Displaced Persons, edited by Carleen F. Maitland

Designing an Internet, David D. Clark

Reluctant Power: Networks, Corporations, and the Struggle for Global Governance in the Early 20th Century, Rita Zajácz

Human Rights in the Age of Platforms, edited by Rikke Frank Jørgensen

The Paradoxes of Network Neutralities, Russell A. Newman

Zoning China: Online Video, Popular Culture, and the State, Luzhou Li

Design Justice: Community-Led Practices to Build the Worlds We Need, Sasha Costanza-Chock

Fake News: Understanding Media and Misinformation in the Digital Age, edited by Melissa Zimdars and Kembrew McLeod

Researching Internet Governance: Methods, Frameworks, Futures, edited by Laura DeNardis, Derrick L. Cogburn, Nanette S. Levinson, and Francesca Musiani

Red Lines: Political Cartoons and the Struggle against Censorship, Cherian George and Sonny Liew

Seeing Human Rights: Video Activism as a Proxy Profession, Sandra Ristovska

Farm Fresh Broadband: The Politics of Rural Connectivity, Christopher Ali

Cyberinsurance Policy: Rethinking Risk in an Age of Ransomware, Computer Fraud, Data Breaches, and Cyberattacks, Josephine Wolff

Resistance to the Current: The Dialectics of Hacking, Johan Söderberg and Maxigas

Universal Access and Its Asymmetries: The Untold Story of the Last 200 Years, Harmeet Sawhney and Hamid R. Ekbia

The Power of Partnership in Open Government: Reconsidering Multistakeholder Governance Reform, Suzanne J. Piotroswki, Daniel Berliner, and Alex Ingrams

Managing Meaning in Ukraine: Information, Communication, and Narration since the Euromaidan Revolution, Göran Bolin and Per Ståhlberg

The Character of Consent: The History of Cookies and the Future of Technology Policy, Meg Leta Jones